*Zillions of Practice Problems*

*Decimals and Percents*

Stanley F. Schmidt, Ph.D.

PDP

Polka Dot Publishing

© 2021 Stanley F. Schmidt
All rights reserved.

ISBN: 978-1-937032-57-9

Printed and bound in the United States of America

Polka Dot Publishing					Reno, Nevada

*To order copies of books in the Life of Fred series,*

*visit our website* PolkaDotPublishing.com

*Questions or comments? Email the author at* lifeoffred@yahoo.com

Third printing

*Zillions of Practice Problems Decimals and Percents* was illustrated by the author with additional clip art furnished under license from Nova Development Corporation, which holds the copyright to that art.

*for Goodness' sake*

or as J.S. Bach—who was never noted for his plain English—often expressed it:

***Ad Majorem Dei Gloriam***
(to the greater glory of God)

## *How This Book Is Organized*

*Life of Fred: Decimals and Percents* has 33 chapters. So does this book.

As you work through each chapter in *Life of Fred: Decimals and Percents* you can do the problems in the corresponding chapter in this book.

Each chapter in this book is divided into two parts.

★ The first part takes each topic and offers a zillion problems.

★ The second part is called the 𝔐ixed 𝔅ag. It consists of a variety of problems from the chapter and review problems from the beginning of the book up to that point.

Please write down your answers before turning to the back of the book to look at my answers. If you just read the questions and then read my answers you will learn very little.

### ELIMINATING TEMPTATION

The solutions and answers are all given in the back half of the book. The first question in this book is numbered "108." The second one is "119." In most ordinary practice books, they are numbered, "1, 2, 3 . . ." which is really silly when you think about it. In those books, when you look up the answer to "1" you might accidentally see the answer to "2" and that would spoil all the fun.

If you happen to spot an error that the author, the publisher, and the printer missed, please let us know with an email to: lifeoffred@yahoo.com

**SPECIAL OFFER**

As a reward, we'll email back to you a list of all the corrections that readers have reported for this book.

# Chapter One
# Number Systems

## First part: Problems from this chapter

108. Fill in the blanks.
     85.79 = ___ tens + ___ ones + ___ tenths + ___ hundredths

119. Fill in the blanks.
     602.1 = 6 _____ + 0 _____ + 2 _____ + 1 _____

222. Fill in the blanks.
     8174. = 8 thousands + 1 hundred + 7 tens + 4 ones
     _____        10×10        ___        1

325. Fill in the blanks.
     5.238 = 5 ones + 2 tenths + 3 hundredths + 8 thousandths
     ___     1/10     _____        _____

480. Joe paid his rent. He paid the landlord

How much was his rent?

# Chapter Two
# Adding Decimals

## First part: Problems from this chapter

130.  $6.4 + 9.7 = ?$

250.  $\$45 + \$2.98 = ?$      (Hint: $45 is the same as $45.00.)

380.  $0.038 + 0.88 = ?$

567.  Add, using decimals.  $\frac{7}{10} + \frac{6}{100} + \frac{3}{100} + \frac{8}{100} + \frac{5}{1000}$

690.  $500 + 0.500 = ?$

720.  $\$9.49 + \$0 = ?$

Chapter Two  Adding Decimals

# Second part: the 𝕸ixed 𝕭ag: a variety of problems from this chapter and previous chapters

150. Darlene and Joe went shopping at the largest store in town.

At the King KITTENS mall, Darlene bought some nail polish ($17.77), some lipstick ($34.99), and some hair spray ($15.95).

Joe bought some fishing line ($16.55), some bait ($8), and some bags of jelly beans ($43.45).

Who spent more money?

240. Fill in the blanks.
    82,507 = ___ ten thousands + ___ thousands + ___ hundreds + ___ tens + ___ ones.

333. In our **base 10 system**,
    6419 = 6 thousands + 4 hundreds + 1 ten + 9 ones, which can be written as 6 × 10×10×10 + 4 × 10×10 + 1 × 10 + 9 × 1

    In the **base 20 system** (also known as the vigesimal system)
    6419 means 6 × 20×20×20 + 4 × 20×20 + 1 × 20 + 9 × 1

    712 in the base 20 system equals what in our base 10 system?

888. In the base 2 system (also known as the binary system)
    1011 means 1 × 2×2×2 + 0 × 2×2 + 1 × 2 + 1 × 1.

    $11101_{base\ 2}$ is equal to what in our base 10 system?

7

## Chapter Three
## Subtracting Decimals

### First part: Problems from this chapter

117.  5.7 – 1.27 = ?

242.  0.611 – 0.4789 = ?

400.  0.7 means the same thing as 0.70.  To show that this is true, change both 0.7 and 0.70 into fractions and show that they are equal.

750.  67.2 means the same thing as 67.200.  To show that this is true, change both into fractions and show that they are equal.

830.  At King KITTENS mall Joe found exactly what he wanted.  It was a backpack with the word "School" on it.  He knew that would make him look smart.

He paid ten dollars for it, which included a sales tax of $0.75.  How much was the price of the backpack before the sales tax was added?

900.  Joe thought to himself, "What should I put in my backpack?"  The answer was obvious.

He filled his backpack with jelly beans from his 100-pound sack. After filling his backpack, his 100-pound sack now weighed 43.7 pounds. How many pounds of jelly beans had Joe transferred?

Chapter Three  Subtracting Decimals

# Second part: the 𝔐ixed 𝔅ag: a variety of problems from this chapter and previous chapters

160. Joe had transferred 87 cherry-marshmallow jelly beans to his backpack. (He had counted them.)
    Express $87_{\text{base 10}}$ in base 2 (the binary number system).

330. Express $87_{\text{base 10}}$ in base 20 (the vigesimal number system).

508. Cherry-marshmallow jelly beans are one of Joe's favorites.

Joe wrote to the jelly bean company and asked for their recipe.
They wrote back to him:

> Dear Joe,
>     Our recipe is a secret. We will tell it to you only if you promise not to tell it to anyone else.
>     Also, please enclose $97.99 for our shipping, handling, and legal expenses.
>
>         Sincerely,
>         *Max*
>         Max A. Million, president
>         Jelly Bean International

Joe had $100 in his checking account. He mailed Max a check for $97.99. How much did he have left?

697. Joe also wrote, "I will keep your recipe a secret. I will never tell anyone."
    Here is what Max sent Joe:

The Secret Cherry-Marshmallow Recipe
▶▶ not to be revealed to anyone ◀◀

To make 70 kilograms . . .
1. 2 grams of cherry flavor
2. 3.8 grams of marshmallow
3. the rest is sugar.

How much sugar?
Hint: A kilogram is 1,000 grams. So 70 kilograms is 70,000 grams.

840. What fraction of a cherry-marshmallow jelly bean is sugar?

# Chapter Four
## Multiplying By Ten

### First part: Problems from this chapter

123. Multiply each of these by ten.
   3.45   9.9   0.002   550

225. Multiply each of these by a thousand.
   2.789   4   7.007   0.0063

339. Fred saw this in the newspaper.

## THE KITTEN Caboodle

The Official Campus Newspaper of KITTENS University         Monday 8: a.m. Edition  10¢

**exclusive**
## Duck Fund Success

KANSAS: "The stock market has been conquered," announced today by the head of Nefarious Investment Corporation.

C.C. Coalback said that his corporation's Duck Fund has multiplied investments by a thousand fold.

"Every investor's dollar has been turned into a thousand dollars," he told a KITTEN CABOODLE reporter this morning.

Coalback urged everyone to send him money to invest in Duck Fund. He said, "Send any amount. Hurry. This is a limited time offer."

Duck Fund Investor

Fred had $2,400 in his checking account. He wrote a check for $2,300 and mailed it to the Duck Fund. (He kept $100 in his account to pay for things until his investment multiplied.)

How much was Fred expecting his $2,300 to turn into?

411. Fred would eventually get back one-hundredth of his $2,300 investment. He was lucky. Most investors in the Duck Fund would get back nothing.

How much did Fred get back?

Chapter Four  Multiplying By Ten

## Second part: the 𝔐ixed 𝔅ag: a variety of problems from this chapter and previous chapters

170. C.C. Coalback originally started out with $23.07. He spent $4.88 investing in all those funds: Aardvark Fund, Alligator Fund, Ants Fund, etc. How much did Coalback have at this point?

280. Many cartoon characters have only eight fingers, instead of ten.
   Instead of the base 10 decimal system, those characters use a base 8 numeration system.
   Fill in the blanks.
   $751_{\text{base }10}$ = __ × 8×8×8 + __ × 8×8 + __ × 8 + __ × 1

398. Add, using decimals.  $\frac{7}{100} + \frac{943}{1000} + \frac{8}{10}$

525. Money poured in for Coalback after that newspaper article. Fred had sent in $2300. Other people had sent in $8.68, $74.99, and $930. How much had these four people sent in?

640. Show that 0.85 and 0.850 are equal by changing them both into fractions and then showing that the fractions are equal.

777. Subtract 0.294 from 0.85.

852. In our base 10 decimal system 731 = 7 × 100 + 3 × 10 + 1 × 1. What does 00731 equal?

## Chapter Five
## Pi

### First part: Problems from this chapter

155. When you were first learning about circles, we named the three parts of a circle: the radius, the diameter, and the circumference.

The easy part: The diameter is twice as long as the radius.

The harder part was finding the circumference, which is the distance around the circle.

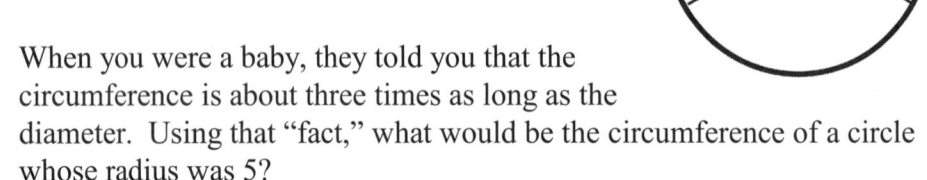

When you were a baby, they told you that the circumference is about three times as long as the diameter. Using that "fact," what would be the circumference of a circle whose radius was 5?

260. After you were potty-trained, they told you that if you wanted a better answer, you multiply the diameter by $3\frac{1}{7}$ to get the circumference of a circle. They said that using 3 was baby stuff.

     Using that "fact," what would be the circumference of a circle whose radius is $\frac{3}{4}$?

470. Again using $C = (3\frac{1}{7}) \times d$

where C is the circumference and d is the diameter, find the circumference of a circle whose diameter is $4\frac{5}{11}$

*This diameter of $4\frac{5}{11}$ looks ugly, but I chose it because it makes the answer come out pretty.*

581. $3\frac{1}{7}$ just gives you a better answer than 3 when you are trying to find the circumference of a circle. To get a better answer you could use 3.1416.

## Chapter Five Pi

In algebra we will write C ≈ 3.1416d where ≈ means "approximately equal to."

So if you had a diameter of 7.02, you could compute the circumference using C = 3.1416 × 7.02.

You are no longer a baby, and you are potty-trained. But you are not old enough to compute 3.1416 × 7.02. You need to be a tiny bit older than you are right now. That's because we are going to learn how to multiply decimals in the next chapter (Chapter 6, Multiplying Decimals).

Please be patient.

In the meantime before you are old enough to multiply decimals, we need to talk about rounding.

Let's start with the number 583.17.

$$5 \quad 8 \quad 3 \,.\, 1 \quad 7$$
hundreds  tens  ones  tenths  hundredths

If we are talking about money, $583.17 makes sense. That's five hundred eighty-three dollars and seventeen cents. (It would be silly to talk about $583.174.)

On the other hand, if you were looking at Fred's trig class and trying to guess how many students were in the giant classroom, you wouldn't say 583.17.

If you wanted to make a rough estimate by just looking, you might round off 583 to the nearest hundred, which is 600.

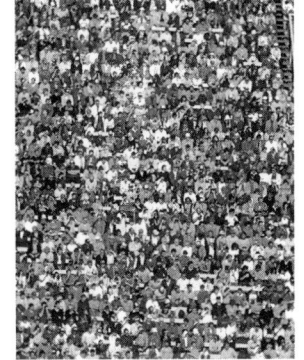

Fred's Trig Class

To round off 583 to the nearest hundred you are getting rid of the 83. 5~~83~~.

Is 583 closer to 500 or closer to 600? ← That's your question to answer.

13

Chapter Five Pi

## Second part: the 𝔐ixed 𝔅ag: a variety of problems from this chapter and previous chapters

230. Many cartoon characters have only eight fingers and eight toes, instead of our 20 fingers and toes.

Instead of the base 10 decimal system, those characters use a base 16 numeration system.

Let's say that the sixteen digits in the base 16 system are 0, 1, 2, 3, 4, 5, 6, 7, 8, 9, a, b, c, d, e, f.

Fill in the blanks.
$751_{\text{base 10}}$ = ___ × 16×16×16 + ___ × 16×16 + ___ × 16 + ___ × 1

336. To round 583.17 to the nearest tenth, you get 583.2 since you are getting rid of a 7. 583.1**7** That is called rounding up. If the digit you are getting rid of is 5, 6, 7, 8, or 9, then you round up.
Examples:
Round 32**8**.9886 to the nearest ten and get 330 since 8 is 5 or more.
Round 0.4**5**03 to the nearest tenth and get 0.5 since 5 is 5 or more.
Round 3.**7**083 to the nearest one and get 4 since 7 is 5 or more.
Round 9**6**2 to the nearest hundred and get 1000 since 6 is 5 or more.

If the digit you are trying to get rid of is 0, 1, 2, 3, or 4, then you just get rid of it instead of rounding up.
Examples:
Round 3**2**8.9886 to the nearest hundred and get 300 since 2 is less than 5.
Round 0.45**0**3 to the nearest hundredth and get 0.45 since 0 is less than 5.
Round 3.7**0**083 to the nearest tenth and get 3.7.
Round 96**2** to the nearest ten and get 960.
Your turn.
Round each of these to the nearest hundred.
673    82.0082    10,501    49

656. Find the sum of 7.8, 84.09, and 0.885. ("Sum" means add.)

716. Fred's trig class has 600 students. Thirty of them use a fountain pen to take notes. What fraction of his class uses a fountain pen?
(The four high school math classes are beginning and advanced algebra, geometry, and trig. Some schools call them alg 1, alg 2, geometry and pre-calculus.)

# Chapter Six
## Multiplying Decimals

### First part: Problems from this chapter

Here's a quick review:  6.73  has 2 digits to the right of the decimal
× 23.8354  has 4 digits to the right of the decimal

So the answer must have 6 (2 + 4) digits to the right of the decimal.

180. 7.8 × 9.7 = ?

390. 69 × 0.007 = ?

436. 0.05 × 0.09 = ?

533. Darlene dreamed that after she married Joe, they would live in a mansion. Their living room would look like this. ⟶

The floor (a rectangle) would measure 20.1' × 45.3'. How many square feet is that?

785. When Joe saw the picture of their mansion's living room, he told Darlene that the dining room would have to be at least 346.5 square feet.
  Darlene said, "We'll make it ten times larger than that. After all, we will want plenty of room to entertain all our friends."
  Joe said, "Betty, Alexander, and Fred?"
  How many square feet would the dining room have using Darlene's plans?

944. Darlene informed Joe that one room of her mansion would be devoted to a kitty litter box. The box would measure 10.8' × 18.'
  "That's a pretty large cat pooping box," Joe exclaimed.
  How many square feet will that be?

## Chapter Six  Multiplying Decimals

## Second part: the 𝔐ixed 𝔅ag: a variety of problems from this chapter and previous chapters

292. You start with the diameter of a circle. The absolutely true length of the circumference of the circle is C = πd.
  π (pi) is the decimal number

  3.14159265358979323846264338332795. . . .

  Pi goes on forever. Pi doesn't repeat itself. In contrast, $\frac{1}{7}$ equals 0.142857142857142857142857142857142857142857. . . .  If you look closely, you can see that $\frac{1}{7}$ does repeat itself.  0.142857142857142857142857142857142857142857. . . .

  In order to use π in everyday life, we have to round it off.

Round off π to the tenths place.

332. Round off π to four decimal digits (to the ten thousandths place).

585. "We'll need horses at our mansion," Darlene told Joe. She showed him a picture. Joe said that he already knew what horses looked like.

horses

"And we will need a nice big circular corral to keep all the horses in," she explained.

If the diameter of Darlene's corral is 1.1 miles, what will the length of the fence for that corral be? (Use 3.1 for π.)

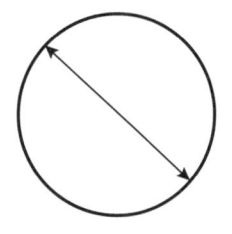

700. (continuing the previous problem) If fencing for the corral costs $1000 per mile, how much will the fencing cost for the whole corral?

810. Joe's pet, whom he'd name Jelly Bean, would eat 4.6 tons of food every 6 days. How much would J.B. eat in 33 days? (Hint: Use a conversion factor.)

# Chapter Seven
# Functions

## First part: Problems from this chapter

A **function** is any fixed rule or procedure. Suppose you are given a list of people: {Fred, Alexander, Betty, Joe, Darlene}. One possible function would be *count the number of letters in each of their first names.*

The answer for each person would always be the same. ← important

Under that rule, Fred would be 4. Alexander would be 9. Betty would be 5. Joe would be 3. Darlene would be 7. Fred wouldn't be 4 one day and 5 the next day. With a function, each member of the starting set always has exactly one answer.

110. Suppose the starting set is {Fred, Alexander, Betty, Joe, Darlene}. Would the rule *Assign each person to the woman who gave birth to each of them* be a function?

203. Suppose the starting set is the set of all people who are currently students at KITTENS University. Would the rule *Assign to each student their art history teacher* be a function?

340. Suppose the domain (the starting set) is the natural numbers, which are {1, 2, 3, 4, 5, 6, 7, . . .}. The rule is *take the number and multiply it by 7.* Is this a function?

456. Suppose the domain is the set of teachers at KITTENS University. Would the rule *Assign to each teacher the day of the week that they were born on* be a function?

528. Suppose the domain is {Fred, Alexander, Betty, Joe, Darlene}. Is the rule *Assign to each member of the domain the color of the jelly bean that they have most recently eaten* a function?

662. Suppose the domain is {✿, ✉, ✈} and the rule is ✿ → Fred and ✉ → Betty.
(Often "→" means "is assigned to.")
Why is this rule not a function?

## Chapter Seven  Functions

759. Suppose the domain is {✿,✉,✈} and the rule is ✿ → A and ✉ → C and ✈ → L and ✉ → R. Is this rule a function?

836. One of Joe's favorite parts of math was functions. In his notes he wrote **All you got to do is be able to count up to one.**

Joe was right. If each member of the starting set (the domain) is assigned to ONE thing, then it's a function.

When Joe went out fishing in his boat, he loved to play the Is-This-a-Function? game.

He liked the phrase "starting set" rather than "domain." The word *domain* seemed too complicated for Joe.

He let the domain (the starting set) be the set of all the fish in the lake that his boat was in.

He let the rule be *Assign the fish to* ☺ *if I catch you today, and assign the fish to* ☹ *if I don't catch you today.* (Joe always thought that fish liked to be caught.)

Is this a function?

915. Is this a function?

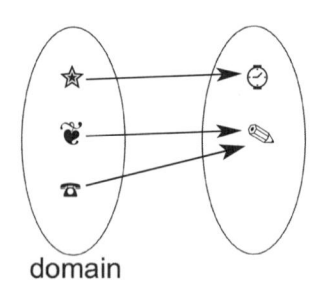

962. Which of these are functions?

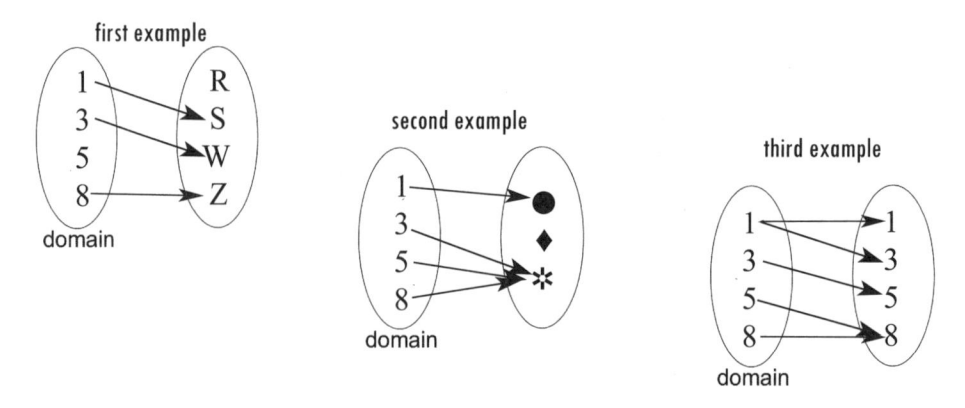

18

Chapter Seven Functions

## Second part: the 𝔐ixed 𝔅ag: a variety of problems from this chapter and previous chapters

246. π = 3.14159265358979323846264338327795....
Round off pi to the nearest hundredth.

327. Joe thought that Jelly Bean would be the perfect pet. If Darlene was going to have zillions of cats and zillions of horses, he wanted to have one animal of his own.

Darlene had seen ads for cats in her Cutie Cat magazine. She had seen ads for horses in her Hunky Horses magazine. Neither of those magazines had what Joe was looking for.

He headed to the magazine section at King KITTENS mall and looked through the candy magazines. Nothing there. He looked through the fishing magazines. Nothing there. He looked through the male bodybuilding magazines and found the perfect pet.

Jelly Bean

**The He-Man Pet!**

**Only one left!**

Ships from Freedonia.

Shipping cost is $3.87 for each kilogram.
Call for current weight.

Joe knew that he would call his pet Jelly Bean because of his big hands. They could hold a lot of jelly beans.

Joe phoned. They told him that the current weight was 1,800 pounds. How many kilograms is that? (one pound ≈ 0.45 kilograms)

440. (continuing the previous problem) How much will it cost Joe to get Jelly Bean shipped to him?

600. We know what the shipping cost is (from the previous problem). On the phone Joe learned that the cost of the He-Man Pet is $8,887. How much is the total that Joe will have to pay including the shipping cost?

678. We know what the total cost of Jelly Bean will be. Joe has 90¢ saved up. (90¢ = $0.90) How much more will he need to save in order to get his pet?

746. Joe currently saves $0 each week. How much would he save in a century?

# Chapter Eight
## Subtracting Mixed Units

### First part: Problems from this chapter

238. The fall semester at KITTENS University is 18 weeks long. When Joe had done 3 days of the fall semester, he asked Darlene how long it was before the end of the semester. Darlene couldn't figure it out and asked Betty. Show Betty's computation.

300. After Joe was 2 weeks and 2 days into the fall semester, he asked Betty how long before the end of the semester. Show Betty's computation.

450. Joe is 5' 6". Jelly Bean is 75' 4". How much taller is Jelly Bean than Joe?

544. Joe sat down to listen to Fred's lecture on fractions. It was an hour lecture. After 7 seconds Joe realized that he hadn't brought any jelly beans to eat during the lecture. Horrors! He would have to go without jelly beans until the end of the lecture. How long would Joe have to suffer?

651. Joe realized that Jelly Bean would require tons of food. Just to get ready he wrote to the company in Freedonia. Instead of writing, "Please send me three tons of food for Jelly Bean," Joe wrote, "Please send me 3 tons of jelly beans." It was the kind of mistake that Joe often made.

When the shipment arrived, Joe was delighted. He thought that both he and his pet liked the same food. Joe ate 5 ounces "just to see if it was any good." How much was left? (Note: 1 ton = 2,000 pounds and 1 pound = 16 ounces.)

Chapter Eight  Subtracting Mixed Units

## Second part: the 𝕸ixed 𝕭ag: a variety of problems from this chapter and previous chapters

114. Suppose the domain (the starting set) was the three tons of jelly beans that Joe received from Freedonia. Those jelly beans were either the 5-gram or the 18-gram varieties. (Five grams is about the weight of a nickel.) They came in red, green, pink, yellow, brown, and blue.
    Joe decided to sort them by the rule *Put the 5-gram jelly beans in one pile, the green ones in a second pile, and all the rest in a third pile.* Why is this not a function?

270. The main room in Joe's studio apartment measures 12.2' by 14.7'. The ceiling is 8.2'. What is the volume of this room? (The volume of a box is equal to length times width times height. In algebra we might write V = 𝓵𝓌𝒽.)

334. (continuing the previous problem) Round that volume off to the nearest cubic foot.

406. Four pounds of Freedonia jelly beans occupied one cubic foot. The dump truck delivered 6,000 pounds (= 3 tons) to Joe's front door. Using a conversion factor, how many cubic feet were in that delivery?

764. Let's say that the sixteen digits in the base 16 system are 0, 1, 2, 3, 4, 5, 6, 7, 8, 9, a, b, c, d, e, f.
    What does b × c equal in the base 16 system? (Hints: First convert b and c to base 10. Then find the answer in base 10. Then convert back to base 16.)

890. Joe's future pet, Jelly Bean, weighs 1,800 pounds. One small jelly bean weighs 0.4 ounces. (1 pound = 16 ounces)
    What is their difference in weight?

977. This is a picture of Joe's old boat. He decided he wanted to make it more like a house. He had a carpenter cut a hole in the side of his boat. Joe thought that would make a nice window.

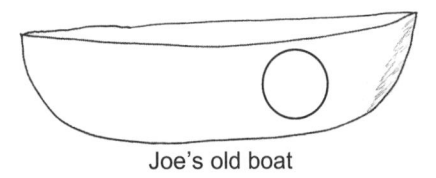

Joe's old boat

21

## Chapter Eight  Subtracting Mixed Units

The hole had a diameter of 2.5 feet.  You can tell from the picture that the boat is only about 3 feet tall.  When Joe was sitting in the boat, he could easily look out over the top of the boat.

Joe was going to ask Darlene to make a nice lace ribbon to go around the hole to make it look pretty.

How long should the ribbon be.  (Use 3.1 for $\pi$.)

This was Joe's *old* boat.  When he put it into the water, it sank.  He had to buy a new boat. He hadn't put any glass in his boat "window."

# Chapter Nine
## Sets

### First part: Problems from this chapter

164. Which of these are true?
    A) 5 ∈ {4, 5, 8}
    B) red ∉ {blue, green}
    C) cow is a member of {cow, horse, star}
    D) duck is not an element of {✽, ✽, ✽}

254. Which of these are true?
    A) California ∈ { x | x is a state in the United States}
    B) 2 ∈ { x | x > 7}
    C) Fred ∉ {x | x is a teacher at KITTENS University}
    D) {15, 16} = {x | x is a natural number and 14 < x < 17}

350. Which of these are subsets of {C, M, V}?
    A) {C}
    B) {M, V}
    C) {V, M, C}
    D) {A, B, C, M, V}

463. Which of these are true?
    A) ∅ = { }
    B) { } ⊂ {2, 18, 444}

513. The union of two sets is abbreviated as ∪. Can you think of a way to remember that?

609. Complete each of these.
    A) {dog, cat} ∪ {cat, mouse} = ?
    B) {Shelly} ∪ {Shelly} = ?
    C) { } ∪ {K} = ?

673. Complete each of these.
    A) {red, orange, blue} ∩ {orange, blue} = ?
    B) {red, orange, blue} − {orange, blue} = ?
    C) {x | x is either a cat or a dog} − {x | x is a cat} = ?
    D) {x | x is an odd number} ∩ {x | x is an even number} = ?

Chapter Nine  Sets

## Second part: the 𝔐ixed 𝔅ag: a variety of problems from this chapter and previous chapters

125. Darlene read in one of her bridal magazines that before she gets married she should walk 3 miles every day. By noon one day she had walked 290 feet. How much farther will she need to walk in order to walk 3 miles? (1 mile = 5,280 feet)

207. Having the right fashionable shoes with which to do her walking was very important to Darlene. She wanted to create a function which assigned to each of her nine pairs of walking shoes to one of the days of the week. The pink pair was assigned to Sunday. The white pair to Monday. The orange pair to Tuesday. The red, to Wednesday. The purple, to Thursday. The blue, to Friday. The brown, to Saturday.
    Was this a function?

306. Joe wanted to assign each of his nine fishing flies to pockets in his new fishing jacket that Darlene gave to him. It has ten pockets.
    Joe stuck all nine fishing flies in the pocket over his heart. He did that because he knew that fishing was close to his heart.
    If the domain is the nine fishing flies, is this a function?

490. Round each of these off to the nearest hundred.
    772    8225    61.0086    5.9

577. This week, Joe's favorite jelly bean flavor is avocado. He weighed eight of those beans (they are all the same size) on the chemistry scale at school. Those eight beans weighed a total of 5.7 grams.
    How much would 56 avocado jelly beans weigh? (Use a conversion factor.)

670. Joe wasn't concentrating. He accidentally put one of his avocado jelly beans on a fishhook. He didn't want to waste his jelly bean so he decided to nibble the bean off of the hook.
    This was a mistake. Instead of catching a fish, he caught himself. Joe rowed to the shore and headed to the doctor's office. The doctor bill was $605. His student health insurance paid $433.64. Joe had to pay the rest. How much was that?

# Chapter Ten
# Rules of Divisibility

## First part: Problems from this chapter

276. Joe owed $171.36 to the doctor because of his fishing accident. Would it be possible for him to make three equal payments?

356. What is the smallest member of this set? {x | x is a whole number that is divisible by 5 and x > 17}

408. Joe had packed 38,394 lemon jelly beans for his fishing trip. Could he divide them equally into five pockets of his new fishing jacket?

519. (continuing the previous problem) Could he divide them equally into two pockets?

738. A) Is 70000000000050000006 divisible by 3?
     B) Is it divisible by 5?
     C) Is it divisible by 2?

Chapter Ten  Rules of Divisibility

## Second part: the 𝔐ixed 𝔅ag: a variety of problems from this chapter and previous chapters

166. Fill in the blanks.
 6.107 = __ tens + __ ones + __ tenths + __ hundredths + __ thousandths

227. To multiply by 10 you move the decimal 1 place to the right.
To multiply by 100 you move the decimal 2 places to the right.
To multiply by 1,000 you move the decimal 3 places to the right.
To multiply by 10,000 you move the decimal 4 places to the right.
 How many places to the right do you move the decimal if you are multiplying by 1,000,000?

360. If a circle has a diameter of a million kilometers, what is its circumference? (Use 3.1416 for $\pi$.)
(A kilometer is a little more than half a mile.)

402. (continuing the previous problem) Round that circumference off to the nearest million kilometers.

540. Suppose A = {4, 5, 6} and B = {5, 6, 88}.
 Find A ∪ B, A ∩ B, and A – B.

643. Joe had a leak in his boat. It was leaking at the rate of 6 liters every four minutes. (He knew that because the water that leaked in 4 minutes would fill his 6-liter Sluice bottle.) If the boat leaked 198 liters, it would sink. (This wasn't the first time that Joe had been in a leaking boat.)

 How many minutes would it take Joe's boat to sink? (Use a conversion factor.)

705. (continuing the previous problem) First, how many hours and minutes will that be? Second, how many hours will that be?

# Chapter Eleven
## Dividing a Decimal by a Whole Number

### First part: Problems from this chapter

191. Darlene's mom could not pay for the wedding that Darlene was planning. Instead, Darlene made up a list of her 17 closest friends. She figured that those friends would be happy to each pay their fair share of the $8,029.61 wedding costs. How much would each owe?

266. Darlene went to several jewelry stores and finally found the perfect engagement ring. She knew that Joe would too busy to come along with her.
    She knew that Joe couldn't afford it in one payment, so she bought it and arranged for the store to have Joe to make 36 easy monthly payments. How much would it cost Joe each month?

$2,257.92

376. Darlene knew that when she finally got Joe to propose to her, he would not have thought of minor things like an engagement ring. She was doing him a favor by doing some thinking for him.

*Author's small note: When I proposed to my future wife, I did not present her with an engagement ring. Instead, we had the fun of our both going shopping together for rings. We found rings that we both liked.*

    Darlene was also helping Joe by planning the mansion that they would live in. In her BRIDAL MANSIONS magazine, she saw the perfect home. It was called a "Starter Mansion"—something perfect for newly weds.

The Perfect Starter Mansion
6 Bedrooms
4816.0 square feet

Darlene thought that it would only be fair if Joe had a little spot to build a shack for himself—maybe one-fortieth ($\frac{1}{40}$) of the size of her mansion. How many square feet would that be?

a spot for his shack

Chapter Eleven  Dividing a Decimal by a Whole Number

## Second part: the 𝔐ixed 𝔅ag: a variety of problems from this chapter and previous chapters

100. Find the first whole number greater than a million that is divisible by both 3 and 5.

199. To go from the diameter of a circle to its circumference you multiply by pi ($\pi$). Pi can be either 3 or $3\frac{1}{7}$ or 3.14 or 3.1416 or 3.1415926535897932384 depending on how much accuracy you want.

    If you want to go from the circumference of a circle to its diameter you do the opposite operation: you divide by $\pi$.

    Darlene found the perfect table for the dining room in her future mansion. It was called the King Arthur Round Table. She walked around the edge of the table and estimated that it had a circumference of 47.1 feet.

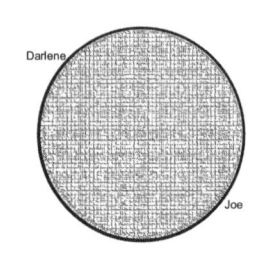

    If she and Joe sat across the table from each other, how far apart would they be?  (Use 3 for $\pi$.)

294. (continuing the previous problem) How far would they be apart if measured in inches.  (Use a conversion factor.)

494. (continuing the previous problem)  If Joe were to take some 8-inch (diameter) plates and line them up between himself and Darlene, how many plates could he fit across the table?

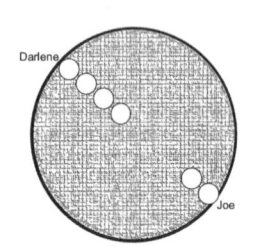

    Hint: The answer won't come out even.  It's almost like Real Life.  If the answer came out, say, 34.8 plates, you would say that he could get at most 34 plates between them.

    Just because you know how to round, doesn't mean that you automatically do it.  In this problem we are looking for a whole number (0, 1, 2, 3, 4. . .) that is ≤ 34.8.  That number is 34.  Thirty-five plates would not fit between Joe and Darlene.

## Chapter Eleven  Dividing a Decimal by a Whole Number

579. Darlene imagined herself sitting at her King Arthur Round Table with Joe sitting 188.4 inches away. He would be busy adding extra salt to his french fries.

In her sweetest voice she could politely ask, "Dearest Joe, please pass the salt if you would."

When Joe wasn't fishing or eating jelly beans, he liked to play with his model trains. He might have built a track between himself and Darlene. It would be 188.4 inches long.

He would put the salt shaker on his train and send it toward Darlene. As it headed toward Darlene, Joe could make the train go "Toot! Toot!"

If it went 3.1 inches per second, would it reach Darlene in a minute?

815. When the salt shaker left Joe's side of the table, it contained 60.2 grams of salt.

On the train it fell over and leaked salt. During the trip it lost 59.48 grams of salt. How much arrived at Darlene's?

## Chapter Twelve
## When Division Doesn't Come Out Even

### First part: Problems from this chapter

121. Change $\frac{5}{8}$ into a decimal. (That's done by dividing 5.00000 by 8.)

209. To go along with her King Arthur Round Table, Darlene knew that she would need forks. In her BRIDAL FORKWARE magazine she spotted the perfect fork. It was made of black steel and featured squared off ends. "No pointy ends. Your guests will not injure themselves." Darlene knew that would be the right kind of fork for Joe.

It was on special: $148.35 for 23 of them. How much would each fork cost?

304. One of the nicest thing about these black steel forks is that they would stack up. Five of them made a stack that was 7 cm (cm = centimeter) tall. How tall was each of them?

421. Those 23 forks cost $170 when you include the shipping, the handling, and the taxes. They could be paid for in "8 easy monthly payments." How much would Darlene pay each month for all these wonderful forks?

620. Even though Darlene didn't own the table
    or own the mansion
        or was married to Joe
            or even engaged to Joe,
                Darlene bought the forks.
She thought that maybe owning the forks would encourage Joe to think about marriage.

She laid them out on her kitchen table in her apartment so that Joe might notice them. After they had lain there for 6 days, she noticed that the first fork had rusted. (That's why you don't hear much about black steel forks.) How long would it take for all 23 forks to rust?

Chapter Twelve  When Division Doesn't Come Out Even

## Second part: the 𝔐ixed 𝔅ag: a variety of problems from this chapter and previous chapters

128. Change 0.25 into a fraction.

386. Which of these are true?

  A) 3 ∈ {x | 1 < x ≤ 7 and x is a whole number}

  B) {7, 10} ⊂ {x | 6 < x < 11 and x is a whole number}

  C) 5.4 ∈ {x | x is a whole number}

  D) $\frac{25}{5}$ ∉ {x | is a natural number}

444. Darlene invited Joe over for dinner at her apartment. It was just going to be a quiet intimate dinner with her, Joe, and her mother.

They were going to share a big bowl of rice. That is all that Darlene could afford. She asked Joe to scoop it out onto three plates. There were 8,200,000 grains of rice. Joe had counted them. Would he be able to give everyone the same number of grains?

516. That afternoon Joe had gone to a JBR (also known as a Jelly Bean Rally). At the rally there was a 218.4 pound jelly bean. It was the largest jelly bean that Joe had ever seen. That giant jelly bean was shared equally by the 600 people at the rally. How much did Joe get?

549. (continuing the previous problem) Using a conversion factor, determine how many ounces of that giant jelly bean Joe received. (1 pound = 16 ounces)

701. (continuing the previous problem) Round your answer off to the nearest ounce.

800. At the dinner party the beverage was Sluice. Joe poured a glass of it on his rice because, as he said, "The rice isn't sweet enough."

Darlene had been at a bridal shower that afternoon (while Joe at been at the JBR). Cake and Sluice had been served. Instead of taking leftover cake in a doggy bag, Darlene had brought a rubber bag and had collected 6 gallons of Sluice.

Seven ounces had leaked out on the way home to her apartment. How much Sluice did she have left for her dinner party?

(1 gallon = 4 quarts.  1 quart = 32 ounces.)

# Chapter Thirteen
## When Division Never Comes Out Even

### First part: Problems from this chapter

184. The JBR (Jelly Bean Rally) also featured potato chips. This was part of their Eat Your Vegetables campaign. None of the people who attended the JBR realized that potatoes are not vegetables.

The main attraction was the one-ton chip. Joe couldn't figure out how they found a potato that large.

A car smashed into it and broke that chip into six equal pieces, each weighing a sixth of a ton.

Express $\frac{1}{6}$ as an unending decimal. (Divide 6 into 1.00000. . . .)

This was called Choice #1 in Life of Fred: Decimals and Percents.

256. (continuing the previous problem) Express $\frac{1}{6}$ as a decimal and express the "leftovers" as a fraction.

This was called Choice #3 in Life of Fred: Decimals and Percents.

341. (continuing the previous problem) Express $\frac{1}{6}$ as a decimal and put a bar over the repeating part.

This was called Choice #4 in Life of Fred: Decimals and Percents.

591. Express $\frac{1}{6}$ as a decimal and round your answer to the nearest thousandth.

This was called Choice #5 in Life of Fred: Decimals and Percents.

647. Expressing $\frac{3}{11}$ as a decimal can be done in several ways.

    A) Express the "leftovers" as a fraction.    Choice #3

    B) Use a bar over the repeating part. Choice #4

    C) Round your answer to the nearest tenth. Choice #5

Chapter Thirteen  When Division Never Comes Out Even

## Second part: the 𝔐ixed 𝔅ag: a variety of problems from this chapter and previous chapters

234. Express one-sixth of a ton of a potato chip in pounds. Round your answer to the nearest pound. (Use a conversion factor. 1 ton = 2,000 lbs.)

452. There are 4 subsets of {M, N}.
They are { }, {M}, {N}, and {M, N}.

    Name the subsets of {2, 9}.

500. Name the 8 subsets of {a, b, c}.

602. At the rally Joe was awarded one of the 333-pound potato chips. It measured 4.9 feet in its largest dimension.
    He decided to build a fence around his chip so that people wouldn't accidentally walk on it and break it. To the nearest tenth of a foot how long would his fence be? (Use 3.1 for π.)

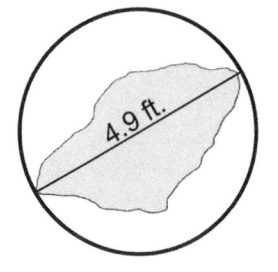

740. Express $\frac{7}{8}$ as a decimal.

844. Express $\frac{312}{999}$ as a decimal.

908. Change 0.33 into a fraction.

# Chapter Fourteen
## Dividing by a Decimal

### First part: Problems from this chapter

102. Darlene was thinking about the mansion that she and Joe would have after they got married. She imagined that the house would be surrounded by a tall brick wall. The wall would be circle with a length of 22.32 miles. What would be the diameter?

(When we knew the diameter of a circle, we could find the circumference by multiplying by $\pi$. When we know the circumference—which in this case is 22.32—we can find the diameter by dividing by $\pi$. Use 3.1 for $\pi$ in this problem.)

162. (continuing the previous problem) If her mansion were at the center of the circle, how far would she have to walk to get from her mansion to the brick wall?

223. She imagined living in that mansion. Every morning she would put on her bathrobe and slippers and walk out to the brick wall to fetch the newspaper. She walks at the rate of 2.4 miles per hour (mph). From the previous problem, you know how far it is (the radius).

Distance is equal to rate times time. d = rt
When we knew the rate and the time, we could find the distance.
But in this case we know the distance (the radius of the circle) and we know the rate (2.4 mph).
When you get to algebra, you will learn to change d = rt into $\frac{d}{r}$ = t. (Right now, that probably looks like magic.)

If you divide the distance by the rate, you will get the time. For example, if you go 180 miles at a speed of 60 mph, it will take you 3 hours. How long will it take Darlene to go out and get the newspaper?

278. Instead of walking out to the brick wall to get the newspaper, Darlene would send her dog, Wolfie. Wolfie could run at 12 mph. How long would it take him to go from the house to the wall?

Chapter Fourteen  Dividing by a Decimal

## Second part: the 𝔐ixed 𝔅ag: a variety of problems from this chapter and previous chapters

126.  If it would take Wolfie 0.3 hours, how many minutes is that? (Use a conversion factor.)

310.  In BRIDAL HOMES magazine Darlene saw a picture of the mansion she wanted. It was on sale for $993,250. She knew that after she and Joe got married they could save $7.25 each week. How many weeks would it take them to buy the mansion? (This can be done with a conversion factor.)

399.  (continuing the previous problem) How many years will that be? (Assume 50 weeks in a year and use a conversion factor.)

536.  Suppose the starting set (the domain) is the set of all 5-bedroom houses within 30 miles of KITTENS University. Would this rule be a function? *Assign to each of those houses either Yes or No depending on whether Darlene and Joe could afford to buy it on their current income.*

611.  With the same domain as the previous problem, Would this rule be a function? *Assign to each of those 5-bedroom houses the number of fireplaces it has.*

819.  With the same domain as the previous problem, Would this rule be a function? *Assign to each of those 5-bedroom houses the color Red if the house has an even number of windows and the color Blue if the house has two bathrooms.*

854.  Is 8000000000000000000005 evenly divisible by 3?

35

# Chapter Fifteen
## Bar Graphs

### First part: Problems from this chapter

168. Walking to class Joe ate 6 cherry jelly beans, 2 lime jelly beans, and 3 chocolate jelly beans. Draw a bar graph to illustrate this.

344. Joe's diet and Fred's diet are very different. Because of that, Joe has 18 cavities in his teeth. Fred has none. Draw a bar graph.

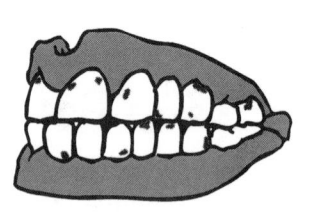

588. Look at these two bar graphs and explain whether Joe or Darlene use more hand lotion each month.

Ounces Used Per Month

Ounces Used Per Month

36

Chapter Fifteen  Bar Graphs

# Second part: the 𝔐ixed 𝔅ag: a variety of problems from this chapter and previous chapters

193.  At the Jelly Bean Rally Joe heard about a new flavor of jelly beans.  They were called Banana Jelly Beans in a Box.  A box cost $1.04.  Joe paid $78.  How many boxes could he get?

378.  If one box of Banana Jelly Beans in a Box cost $1.04, how much would a million boxes cost?

416.  A box of banana jelly beans weighs 5 pounds 3 ounces.  (1 lb. = 16 oz.)  Joe opened a box and ate 7 ounces.  How much did that box now weigh?

521.  A banana jelly bean box measures 8.3" by 7.9" by 10".  What is its volume?  (In algebra we will write V = 𝓵𝓌𝒽.)

663.  (continuing the previous problem)  Round your answer off to the nearest cubic inch.

709.  What is another name for this set?  {x | x is a banana jelly bean that came in one of the boxes that Joe bought and x weighs 14,000 tons.}

860.  Banana jelly beans are just like regular bananas.  They come in three colors: green, yellow, and brown.  Each box contains all three colors.

   Joe opened one of his boxes of banana jelly beans.  Let those beans be the domain (the starting set).  Is this rule a function?  *Joe put the green beans in the refrigerator.  He put the brown beans in the garbage.*

# Chapter Sixteen
# Prime Numbers

## First part: Problems from this chapter

107. None of these numbers are prime numbers.
    400002, 9067097930, 555455554, 7751258, 33388878878878
    What is an easy way to show that they are all composite?

205. All of these numbers are composite.
    60000000003, 70000011, 8870000001, 45
    What is an easy way to show that none of them are prime?

302. Here is a number with the last digit missing.
    56468681168761684684484651611687465117777993?
    What could that last digit be in order to easily know that the number is not prime?

419. Joe filled 11 lunch bags with jelly beans.  Each bag contained 807 jelly beans.

He took them all to Fred's class and ate them all while Fred was lecturing.  A number is composite if it has three or more divisors.  The total number of jelly beans that Joe ate is composite.  Name three numbers that divide evenly into the number of jelly beans Joe consumed.

502. Which one of these is composite?
    57, 71, or 101

Chapter Sixteen  Prime Numbers

## Second part: the 𝔐ixed 𝔅ag: a variety of problems from this chapter and previous chapters

212. Using the base 8 numeration system, fill in the blanks.
$613_{\text{base 10}}$ = __ × 8×8×8 + __ × 8×8 + __ × 8 + __ × 1

258. $400_{\text{base 10}}$ = $?_{\text{base 20}}$

454. π = 3.14159265358979323846264338327 95. . . .
Round π off to the nearest thousandths place.

547. Suppose the starting set (the domain) is the set of all pairs of whole numbers.  For example, (3, 88) or (554, 0) or (209, 2).
Is this rule a function?  *Assign to each pair of whole numbers their sum.*  For example (3, 88) would be assigned to 91.

680. Suppose the starting set (the domain) is the set of all pairs of natural numbers.  For example, (3, 88) or (554, 7777777777) or (209, 2).
The natural numbers are 1, 2, 3, 4, 5, 6. . . .
Is this rule a function?  *Assign to each pair of natural numbers the answer when you divide the first number by the second.*  For example (3, 88) would be assigned to $\frac{3}{88}$.

727. Suppose the starting set (the domain) is the set of all pairs of whole numbers.  For example, (3, 88) or (554, 0) or (209, 2).
The whole numbers are 0, 1, 2, 3, 4, 5, 6. . . .
Is this rule a function?  *Assign to each pair of whole numbers the answer when you divide the first number by the second.*  For example (3, 88) would be assigned to $\frac{3}{88}$.

858. Fill in one word: {4, 7} is a _____ of {4, 5, 7, 8, 9}.

930. Darlene read in BRIDAL DOGS magazine that owning 5 dogs would cost $17.45 per week to feed them.  Darlene thought it would be more fun to have 6 dogs.  How much would it cost to feed 6 dogs?

$17.45

# Chapter Seventeen
# Goldbach Conjecture

## First part: Problems from this chapter

Goldbach's first conjecture was that every even number greater than 2 could be written as the sum of two primes.

### Handy List of Primes
Less Than 1,000

2, 3, 5, 7, 11, 13, 17, 19, 23, 29, 31, 37, 41, 43, 47, 53, 59, 61, 67, 71, 73, 79, 83, 89, 97, 101, 103, 107, 109, 113, 127, 131, 137, 139, 149, 151, 157, 163, 167, 173, 179, 181, 191, 193, 197, 199, 211, 223, 227, 229, 233, 239, 241, 251, 257, 263, 269, 271, 277, 281, 283, 293, 307, 311, 313, 317, 331, 337, 347, 349, 353, 359, 367, 373, 379, 383, 389, 397, 401, 409, 419, 421, 431, 433, 439, 443, 449, 457, 461, 463, 467, 479, 487, 491, 499, 503, 509, 521, 523, 541, 547, 557, 563, 569, 571, 577, 587, 593, 599, 601, 607, 613, 617, 619, 631, 641, 643, 647, 653, 659, 661, 673, 677, 683, 691, 701, 709, 719, 727, 733, 739, 743, 751, 757, 761, 769, 773, 787, 797, 809, 811, 821, 823, 827, 829, 839, 853, 857, 859, 863, 877, 881, 883, 887, 907, 911, 919, 929, 937, 941, 947, 953, 967, 971, 977, 983, 991, 997

104. Express 466 as the sum of two primes.

215. Express 890 as the sum of two primes.

308. Express 270 as the sum of two primes.

  When Christian Goldbach wrote to Leonard Euler in 1742, one obvious question is: Were pencils in existence back then? Goldbach must have done a lot of thinking and writing to invent his famous conjecture. Writing in ink has the drawback of not being erasable.
  The year **1742** is just a number. Let's locate that in history.

**1726** Jonathan Swift wrote *Gulliver's Travels*.
**1727** Quakers demanded the abolition of slavery.
**1728** Danish explorer Vitus Bering and his 33 men sailed north to find out where Russia touched Alaska. Oops. There is no land between Russia and Alaska. They sailed right through.
  That narrow passage of water (known as a strait) is now called the Bering Strait.
**1742** Goldbach wrote to Euler.

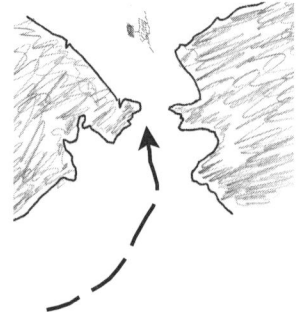

## Chapter Seventeen  Goldbach Conjecture

**1752** Benjamin Franklin flies a kite in a thunderstorm. Luckily, he isn't killed by lightning. (Nowadays, in the United States about 50 people die each year from lightning strikes.)

Okay. We were wondering whether Goldbach could have been using a pencil in 1742 to play with the question of whether every even number, except 2, could be expressed as the sum of two primes.

## pencils

Before or after 1742?

In the 1500s in Italy they were making pencils by drilling a hole in a stick and stuffing in a piece of graphite. Graphite is a mineral. You mine it. It's too brittle to use by itself. (And, besides, your hands would get dirty.) Sticking the graphite inside a hollowed-out wooden stick works much better.

Drilling holes in sticks and slipping in the graphite is tough to do. Later in the 1500s, they found an easier way: make two halves of wood, drop the graphite in and glue the pieces together. That's the way most pencils are made today.

Pencils were first mass-produced in 1662.

The lead in pencils is a mixture of graphite and clay. The more graphite and less clay, the darker the line.

The lead in pencils is not lead. Graphite/clay and lead both are dark gray. Lead is a metal. If pencils contained the metal lead, they wouldn't write well at all.

Don't correct people who talk about lead in their pencils. It's not polite. Even well-educated scientists talk about pencil lead. It's just an expression. When your mom tells you to make your bed, don't get out a hammer and nails! It's just an expression. Instead, straighten the sheets on your bed.

41

Chapter Seventeen  Goldbach Conjecture

# Second part: the 𝔐ixed 𝔅ag: a variety of problems from this chapter and previous chapters

Goldbach's second conjecture was that every number greater than 5 can be written as the sum of three primes.

134. Express 204 as the sum of three primes.
The Handy List of Primes (two pages ago) may help.

228. If three primes add up to an even number, must one of the primes be equal to 2?

312. Express 250 as the sum of three primes.

473. Darlene has been working for years to get Joe to commit to marriage. He doesn't like to commit to anything. For example, he rarely uses a pen. From Joe's point of view, pencils are great. You can always erase.
    If Joe writes 33 words, he will usually erase 7 of them.
    If he writes 264 words, how many will he erase?
    (Hint: Use a conversion factor. We know 33 written words = 7 erased words.)

586. Joe loved to sharpen pencils. He thought that was a lot of fun. He would sharpen and sharpen and sharpen.
    A new pencil would look like this after Joe sharpened it once.
    A new pencil might weigh 20 grams. After Joe got through sharpening it, the pencil might weigh 1.07 grams. How much was lost?

613. After Joe got done sharpening a pencil, he would get so excited that he would write and press too hard. That would break the lead. (We know that's a mixture of graphite and clay.)
    He would throw that pencil away because it was too short to sharpen again.
    At the store he would buy twelve dozen pencils at a time. Twelve dozen (12 × 12) is called a gross. Sometimes he would take that gross of pencils and sharpen them all at once. It would take him 222 minutes.
    A) Convert that into hours and minutes.
    B) Convert 222 minutes into hours.

# Chapter Eighteen
# Area of a Circle

## First part: Problems from this chapter

In algebra r × r is written as $r^2$.
r × r × r is $r^3$.
r × r × r × r is $r^4$.
r × r × r × r × r × r × r × r × r × r × r × r × r × r is $r^{14}$.
    It is all very logical.

    In algebra, they don't use the times sign (×). It gets confused with the letter x, which they use a lot.
    To indicate the multiplication of two letters, they just write them next to each other.
rr = $r^2$
rrrrrrrrrrrrrr = $r^{14}$
    The area of a circle is A = πrr or A = $πr^2$.

157. Darlene was still dreaming about getting a mansion after she married Joe. "We should get some sheep," she told Joe. "Then you can cut off their wool and give it to me. I'll turn the wool into thread and make all of our clothes."

Joe couldn't figure out how Darlene was going to make his rubber flip-flops.

Darlene continued, "We will need a nice big circular corral to keep all my sheep in."

    Darlene had talked about a nice big circular corral to keep her horses in. She was now talking about a nice big circular corral to keep her sheep in. Joe wondered if she was going to get a nice big circular corral to keep him in!

    "My sheep corral will have a diameter of 1.6 miles. That means its radius will be 0.8 miles. I need to know the area of my sheep corral so that I can buy grass seed."
    Find the area. Use 3.1 for π. Round your answer to the nearest tenth of a square mile.

Chapter Eighteen  Area of a Circle

## Second part: the 𝔐ixed 𝔅ag: a variety of problems from this chapter and previous chapters

274. The reason Darlene needed to know the area of her sheep corral was that she needed to buy grass seed. At the store she looked at bags of grass seed.

    How many bags would she need to cover the 2.0 square miles of her sheep corral?

    (Use a conversion factor. You are converting square miles into bags.)

*Grass Seed — This bag covers 0.04 square miles.*

346. Darlene was going to start out simply with just one sheep. In her BRIDAL SHEEP OWNERS magazine she read that the costs would be:

    purchase cost       $179.
    tax                       14.32
    wool cutting scissors    38.55  ← That's not what they are really called. The
    vet costs               60.      magazine article was written by someone
                                                 who had never owned a sheep.

    According to that magazine article, how much would this sheep cost?

425. In BRIDAL SHEEP OWNERS magazine she read an ad:

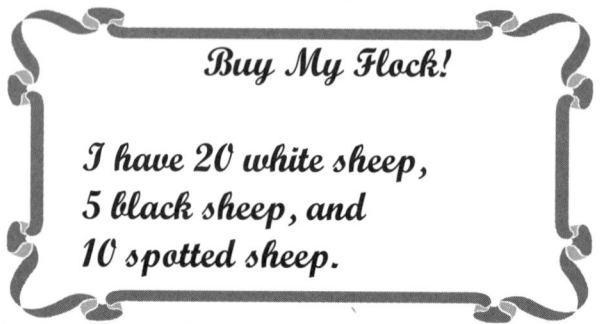

*Buy My Flock!*

*I have 20 white sheep,
5 black sheep, and
10 spotted sheep.*

Draw a bar graph.

504. Darlene thought, "If I had 35 sheep, I would need someone to take care of them. A shepherd. But shepherds cost money. But if Joe marries me, he can be the shepherd for free."

    She asked Joe if he liked sheep.

    Joe answered, "I don't know. Do they taste good?"

## Chapter Eighteen  Area of a Circle

Darlene explained to him that they would be her pets. She would be raising them for their wool, not for mutton. Joe didn't know what *mutton* was. (Mutton = meat from adult sheep)

She told him about her dreams of a sheep corral with a diameter of 1.6 miles. She said that he would be taking care of the sheep.

Joe didn't like that idea. "When could I go fishing, if I have to mess with sheep all the time?"

They reached a compromise. Inside Darlene's sheep corral, they would dig a circular lake that was 0.01 miles across. Joe could put his boat in that lake and fish and also look at the sheep.

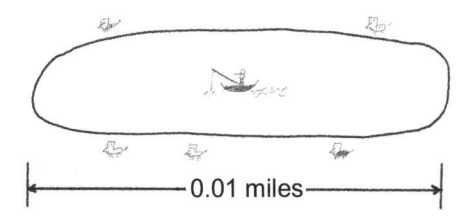

If a sheep walked around that lake, how far would it walk? (Use 3.1 for $\pi$.)

630. How many feet would that be?
(1 mile = 5,280 feet)

711. Round that to the nearest foot.

806. A sheep decided to run around that lake. It can run at 8 feet per second. Use the rounded off number from the previous problem and find how long it will take to circle the lake.
  A) Express your answer as a decimal.
  B) Express your answer as a mixed number.
(Hint: We know that distance equals rate times time.  d = rt  When we get to algebra, we will divide both sides of that equation by r and get $\frac{d}{r}$ = t.)

# Chapter Nineteen
## Dollars vs. Cents

### First part: Problems from this chapter

186. Change into cents.
     $0.25        $3.95        $0.06

264. Change into dollars.
     55¢          744¢         3¢

365. Which of these are probably written by someone confusing $ with ¢?

**Running Into a Brick Wall?**
Eggs!
On sale!
255¢
for a dozen

**Running Into a Brick Wall?**
Pet Mice!
On sale!
.87¢
each

**Running Into a Brick Wall?**
Jelly Beans!
On sale!
.01¢

Chapter Nineteen  Dollars vs. Cents

## Second part: the 𝔐ixed 𝔅ag: a variety of problems from this chapter and previous chapters

132. Fill in the blanks as we change base 10 into base 2.

$28_{\text{base 10}} = \underline{\;?\;} \times 2^4 + \underline{\;?\;} \times 2^3 + \underline{\;?\;} \times 2^2 + \underline{\;?\;} \times 2 + \underline{\;?\;} \times 1$
$\phantom{28_{\text{base 10}} = } \phantom{\underline{\;?\;}} \; 16 \phantom{\times 2^4 +} \phantom{\underline{\;?\;}} \; 8 \phantom{\times 2^3 +} \phantom{\underline{\;?\;}} \; 4 \phantom{\times 2^2 +} \phantom{\underline{\;?\;}} \; 2 \phantom{\times 2 +} \phantom{\underline{\;?\;}} \; 1$

247. $46_{\text{base 10}} = ?_{\text{base 2}}$

393. Darlene told Joe, "You got to vary your diet a little. Eating jelly beans all the time is too much sugar."

Joe found the perfect solution. Everyone knows that soup is good for you.

The width of the can is 3". Find the area of the top of the can. (Use 3.1 for π.) Round your answer off to the nearest square inch.

446. The volume of a can is equal to the area of the top times the height of the can. (In algebra we will write $V = \pi r^2 h$.)

The can is 6 inches tall. (h = 6".)
Its radius is 1.5". (r = 1.5".)
Using the answer to the previous problem, find the volume of the can.

531. From the previous problem we know the volume of a can of Jelly Bean Soup in cubic inches. Round that answer off to the nearest cubic inch.

658. Using the rounded off answer from the previous problem, find how many quarts is a can of Jelly Bean Soup. (It will help to know that a cubic inch is approximately equal to 0.017 quarts.)

> In geometry the shape of a can will be called a cylinder. (SILL-in-der)
> In geometry the shape of a ball will be called a sphere.
> In geometry the shape of a triangle will be called a triangle. ☺

47

## Chapter Twenty
## Pie Charts

### First part: Problems from this chapter

If we know that Fred is three feet tall and Alexander is six feet tall, then that could be expressed in a bar graph.

On the other hand, if we know that the president of KITTENS University makes 70% of the total salaries and the teachers make 25% of the total salaries and the secretaries make 5% of the total salaries, then that can be expressed in a pie chart.

$$70\% \text{ means } \frac{70}{100} \qquad 25\% \text{ means } \frac{25}{100} \qquad 5\% \text{ means } \frac{5}{100}$$

105. Draw a bar graph showing Fred and Alexander's heights.

136. Draw a circle graph (pie chart) showing the salaries at KITTENS.

210. Change to fractions. (Remember to reduce fractions as much as possible.)   37%   50%   1%   100%

315. Change to a percent.
    0.89   0.5   0.04

404. Change to a decimal.
    71%   6%   3.02%

553. Change $\frac{3}{8}$ to a percent. (The usual way to do this is to first change the fraction into a decimal.)

645. One month Darlene spent one-eighth of her income on nail polish, one-eighth on lipstick, one quarter on her hair, and one half on clothes. Draw a pie chart to illustrate this.

Chapter Twenty  Pie Charts

## Second part: the 𝔐ixed 𝔅ag: a variety of problems from this chapter and previous chapters

140. Darlene was going to buy a little grass seed to see how it would look in her sheep corral. She made a mistake in her arithmetic. A big mistake. A zillion trucks arrived at her front door.

One of the drivers told her that she had enough seed to plant a circle with an area of 75 square miles. What would be the radius of that circle? (Use 3 for $\pi$.)

298. (continuing the previous problem) It is difficult to buy a circular piece of land. Suppose she bought a square piece of land that was just big enough to contain that circle of grass. What would be the length of one side of that square?

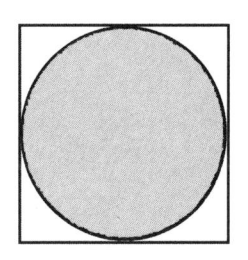

352. *Take a number, double it, and then subtract 4* is a function.
What is the inverse of that function? (Hint: It is not: *Divide by 2 and then add 4*.)

484. The Jelly Bean Soup manufacturers are coming out with two new products.
 ✔ Syrup Soup. It comes in a 9" tall can with a radius of 2".
 ✔ Sugar Soup. It comes in a 4" tall can with a radius of 3".

Syrup Soup                 Sugar Soup

Which can has a larger volume?  ($V = \pi r^2 h$)
(𝔐ixed 𝔅ag continued on next page)

49

## Chapter Twenty  Pie Charts

606.  Syrup Soup is 98% sugar.  Express that as a fraction (reduced to lowest terms).

731.  Syrup Soup is 98% sugar and 2% food coloring.  (It is a lovely blue.)  Draw a pie chart.

838.  Joe prefers Sugar Soup over Syrup Soup.  He thinks Sugar Soup has a better flavor.  (Sugar Soup is 99% sugar and 1% water.)

     He bought a case of Sugar Soup.  The case measured 12" by 12" by 10".  What is its volume?  (In algebra we will write $V_{box} = \ell wh$.)

*Sugar Soup*

Not approved by the American Dental Association

---

### Intermission

    In the old days (before you were born) one of the biggest fears was not getting enough food.  People died of starvation.

    The key to staying alive was finding things that were Sweet and Greasy.

    Fruits that were sweet were ready to eat.

    Meat that was greasy had a lot of fat.  Fat has a lot of calories.  Calories are food energy.

    Our ancestors didn't starve if they found sweet and greasy things.  Their descendants, including Joe, still look for sweet and greasy.

---

    Actually, it was sweet, greasy, and salty.  Salt is hard to find in a forest.  Now it all makes sense: a hamburger, salty french fries, and a milk shake.

# Chapter Twenty-one
## 40% of 15

### First part: Problems from this chapter

*of* means multiply      32% of 25 means 0.32 × 25.

174. KITTENS University pays a million dollars a year in salaries. The president makes 70% of the total salaries paid. How much does he get each year?

220. KITTENS University income is two million dollars a year. It receives 98% of its income from student tuition. (The rest is received from gifts. It receives no funds from the federal or state governments.)
    How many dollars does it receive from tuition?

383. (continuing the previous problem) If there are 5,000 students at KITTENS, what is the tuition that each student pays?

496. Fred's salary is $500/month. He donates 10% of his salary to the Sunday school offering. How much is that?

517. On June 1, Joe had 4,567 pounds of jelly beans stored at his apartment. By the end of the month he had eaten 100% of them. How many pounds had he eaten?

685. Fred is 5½ years old. Let's call it 66 months. He has been teaching at KITTENS University for about 86% of his life. How many months has he been teaching? Round your answer to the nearest month.

$$\frac{5½ \text{ years}}{1} \times \frac{12 \text{ months}}{1 \text{ year}} = 5½ \times 12 \text{ months}$$
$$= \frac{11}{2} \times \frac{12}{1}$$
$$= \frac{11}{\underset{1}{\cancel{2}}} \times \frac{\cancel{12}^{6}}{1}$$
$$= 66 \text{ months}$$

51

Chapter Twenty-one  40% of 15

## Second part: the 𝕸ixed 𝕭ag: a variety of problems from this chapter and previous chapters

137. Darlene saw an ad for a square plot of land.
    A) How many square miles is that?
    B) What is the length of the side of that square?
    (Hint: 640 acres = 1 square mile)

10,240 acres

244. At $1,000 per acre, how much would those 10,240 acres cost?

427. What is the radius of the biggest circle that could fit inside of the square plot of land?

590. What is the area of that circle?  (Use 3.1 for $\pi$.)

665. Darlene had spent 4 hours doing three things: reading the ad, painting her nails, and watching television.
    Painting her nails took 1 hour and 40 minutes.
    Watching television took 2 hours.
    How long did she spend reading the ad?

794. Darlene considered those four hours as her "work day." The rest of the time she spent goofing off.
    Four hours is what fraction of 24 hours?

870. Express $\frac{1}{6}$    A) as a decimal
                     B) as a percent

# Chapter Twenty-two
# 30% Off

## First part: Problems from this chapter

153. After Darlene got married to Joe, she knew that she would no longer have to take the bus. She read in CARS FOR BRIDES magazine that she should expect a new car every year.

She had already picked out her first new car. It was normally $80,000, but it was on sale for 20% off of that price. How much would Joe have to pay for it?

282. Darlene imagined that Joe would get a job and make $1,875/week. He would work Monday through Friday. On Saturday he would mow the lawn at their new mansion and pay the bills. On Sunday they would head to church in the morning and get ready for a big party at their mansion every Sunday evening.

When she told this to Joe, he frowned and said, "Work five days a week? Monday, Tuesday, Wednesday, Thursday, and Friday? I couldn't work that many days in a row. I would need to take Thursdays off so that I could go fishing."

That would cut one-fifth off of his salary. How much would he then be making each week?

small essay
### Making Money

There are four ways to make money: inherit it, marry someone rich, steal it, or earn it.

**Inherit.** Neither Joe nor Darlene have any rich parents. That won't work.

**Marry.** Darlene won't get rich marrying Joe. She can dream about his having enough money for her to buy a mansion and a new car each year, but that is not reality.

**Steal.** The most common way that is done nowadays is to have the government take/steal/tax money from those who have earned it and give it to those who haven't earned it.

## Chapter Twenty-two  30% Off

**Earn.** To make money you need to trade your labor for other people's money. They must want your labor more than their money. That's what doing a trade means.

A brain surgeon offers something that few people can offer. That surgeon has spent years learning that skill and years in practicing it. Patients happily give hundreds of dollars per hour for his work.

What does Joe have to offer other people? He could sweep floors. He could wash dishes. He could clean windows. None of these require much education. Lots of people can sweep, wash, and clean. Joe doesn't have anything special to offer other people. Why should they want to trade lots of their money for his labor?

Joe will probably never make $1,875/week.

end of small essay

321. Joe was trying to sell a fish he had caught.

He set the price at $1.50, but no one wanted to buy it.

Then he put it on sale at two-fifths off the original price. The fish was no longer fresh.

When the first began to rot, he knocked an additional one-third off the sale price.

What was the price after these two discounts were made?

(No one wanted to buy his rotting fish, so he threw it in the garbage.)

Have you ever wondered why kids don't make as much money as grownups? The answer is simple. They don't have as much to offer in trade for their labor.

★ Kids are not as strong. They can't shovel dirt as fast as a man or woman can.

★ Kids can't concentrate as long as a grownup. They might work for 10 minutes and then want to go off and play.

★ Kids don't have the education and experience that grownups have. They can't teach algebra or repair a car or program a computer like someone who has learned those skills.

Chapter Twenty-two  30% Off

## Second part: the 𝔐ixed 𝔅ag: a variety of problems from this chapter and previous chapters

337. What is the smallest member of this set? {x | x is a whole number that is divisible by 3 and x > 10,040,200}

458. Complete each of these.
    A) {moose, rat, fish} ∩ {rat, 6, A} = ?
    B) {moose, rat, fish} − {rat, 6, A} = ?
    C) {x | x is an animal} ∪ {x | x is a dog} = ?
    D) {x | x is a star in the sky} ∩ ∅ = ?

538. Express $\frac{1}{12}$ as a decimal and round your answer to the nearest hundredth.

615. Everyone went out for pizza at Stanthony's PieOne. Three of them had dancing pizza dreams that night. Darlene dreamed about dancing pizzas for 6 minutes. Fred dreamed about dancing pizzas for 12 minutes. Joe, for 3 minutes.
    Draw a bar graph.

Dance!

761. The times that they dreamed were 6, 12, and 3. Which of these numbers are composite?

813. If you add two prime numbers, will the answer always be composite?
    You have probably never been told the answer to this question. It is not a matter of recalling a memory. When you were little, you just memorized stuff. When you were asked, "What is 6 × 9?" you recited from memory, "54."
    As you get older, the questions involve less memorizing and more thinking.

896. Take a peek at the Handy List of Primes (page 40). One of three things is true:
    I) Primes get more plentiful as you go to larger and larger numbers.
    II) Primes seem to be equally scattered among all the numbers.
    III) Primes seem to get rarer as you go to larger and larger numbers.
    Which of these appears to be true? More thinking!

# Chapter Twenty-three
## Distance = Rate × Time

### First part: Problems from this chapter

In algebra we write distance = rate × time as d = rt.

Back on page 34, we noted that $\frac{d}{r} = t$.

158. When chased, a dancing pizza can run 40 mph (miles per hour). How far can it run in 3.2 hours?

261. Express 3.2 hours as hours and minutes.

395. A dancing pizza ran 128 miles. How long would it take Joe to run 128 miles? Joe can run at the rate of 8 mph.
  (Hint: Use $\frac{d}{r} = t$.)

475. While fishing, Joe eats jelly beans at the rate of 27 per minute. How many will he eat in 3 hours? (First, change hours into minutes.)

557. Joe once wondered how long it would take him to eat a million jelly beans if he were eating them at the rate of 27 per minute. Round your answer off to the nearest minute.

677. (continuing the previous problem) How many hours would it take Joe? Again, round your answer off to the nearest hour.

714. (continuing the previous problem) If Joe "worked" at eating jelly beans for eight hours each day, how many days would it take him to eat a million jelly beans? Again, round your answer to the nearest day.

Chapter Twenty-three  Distance = Rate × Time

## Second part: the 𝔐ixed 𝔅ag: a variety of problems from this chapter and previous chapters

139. Darlene imagined that after she married Joe she could retire and just play all day in her mansion. When Joe came home from work at 6 p.m., she would make sure that dinner was ready for him. Their cook would prepare a nice plate of jelly beans for Joe.

The diameter of the plate is 12". What is its area?

236. Joe would get a larger plate—one with an area of 200 square inches. If the cook filled that big plate with jelly beans, each square inch would hold 7.8 jelly beans. How much would the whole plate hold?

317. Darlene imagined that when Joe got home from work, their butler would exchange Joe's shoes for slippers and give him the newspaper comics to read before dinner.

 Joe would spent 50% of his time reading the Donald Duck comic, 40% of his time reading the Woody Woodpecker comic, and 10% of his time resting his eyes. Draw a circle graph (a pie chart) showing this.

410. Darlene imagined that once she was married she would spend much of her time doing something significant. She couldn't think of what that would be.

 Finally, she decided that she would collect vases. She could fill her mansion with vases from around the world.

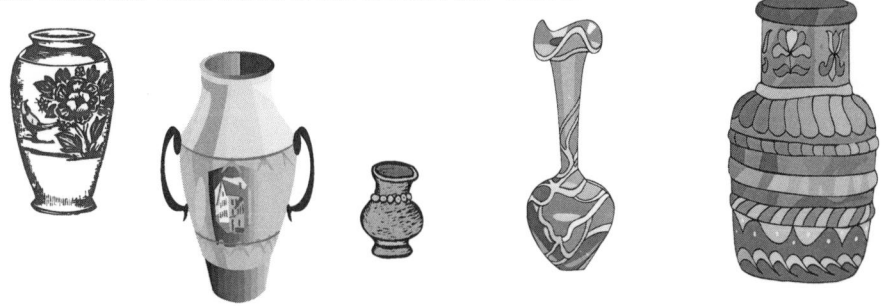

(continued next page)

57

Chapter Twenty-three  Distance = Rate × Time

She would spend her days buying vases on the Internet.  In her first month of collecting, she would buy 400 vases.  Thirty percent of them would be porcelain.  How many would that be?

583.  Darlene saw the vase that she just couldn't live without.  She had to have it.  It was listed at $10,000, but it was on sale for "Five percent off."  She was sure that Joe would love it.

How much would it cost?

# Chapter Twenty-four
## 15% More

**First part: Problems from this chapter**

142. Darlene was delighted with the thought of filling her mansion with vases. In her first month of collecting, when she was just starting out, she would buy 400 vases. In the second month she would have learned how to really buy vases and would buy 34% more than she did in the first month. How many vases did she imagine buying in the second month?

216. When Darlene told Joe about her plans to buy a zillion vases for their mansion that they would own after they got married, Joe didn't think about the cost of a mansion or the cost of all those vases. His mind had wandered off to what he was going to have for lunch that day.

The only thing that stuck in his mind was having a zillion vases. He wondered: *What do you do with a zillion vases?*

The answer was obvious. You fill them with jelly beans! Instead of his usual monthly order of 68 cases of jelly beans, he would order 25% more. How many cases would he order?

362. C.C. Coalback ran an ad in BRIDAL VASES magazine.

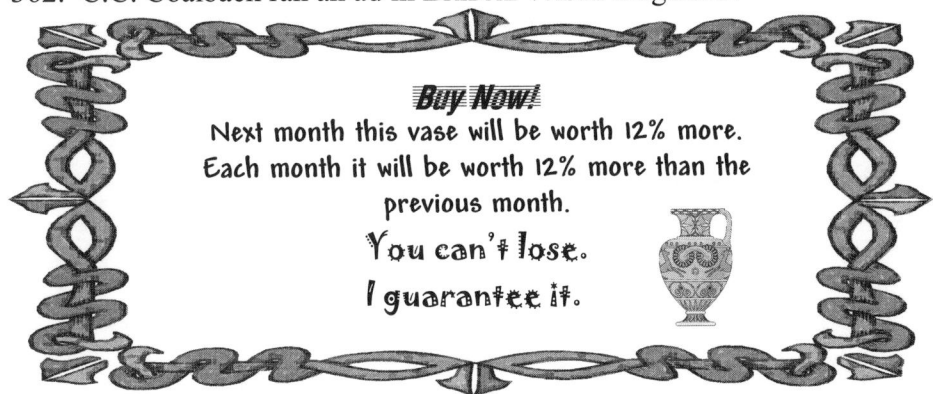

> **Buy Now!**
> Next month this vase will be worth 12% more.
> Each month it will be worth 12% more than the previous month.
> You can't lose.
> I guarantee it.

The vase was only $10. Darlene sent in her money in August. How much would it be worth in September? In October?

Chapter Twenty-four  15% More

## Second part: the 𝔐ixed 𝔅ag: a variety of problems from this chapter and previous chapters

144. Let T = {x | x is a vase and x weighs less than three pounds}
Let F = {x | x is a vase and x weighs less than five pounds}
    A) Is T ⊂ F true?
    B) T ∪ F = ?
    C) T ∩ F = ?
    D) T – F = ?

388. There are three different percent problems that we have done so far. They all involve multiplying. Three chapters from now, we will have the last type of percent problem. It will involve dividing.

    Joe filled one of Darlene's vases with 60 pounds of jelly beans.

Three different problems . . .

    A) 7% of those beans were orange. How many pounds of orange jelly beans were in that vase? (Chapter 21)

    B) If mice ate 7% of all the jelly beans, what would be the weight of the beans that remained? (Chapter 22)

    C) If Joe added 7% more to the original 60 pounds, what would be the new weight? (Chapter 24)

506. Joe weighed 170 pounds. He went on a diet (he stopped eating for ten minutes) and lost 10% of his weight. He looked in a mirror and thought that he was starting to look a little too skinny.

He rushed to the nearest vending machines and emptied them and filled himself. He increased his weight by 10%.

How much did he now weigh after losing 10% and then gaining 10%?

# Chapter Twenty-five
## Area of a Triangle

### First part: Problems from this chapter

141. Darlene wanted her mansion to be special. When she drew plans for the house she made one of the rooms in the shape of a triangle. "No one else has a triangular room," she thought.

She needed to find out how much carpet she would need. She made these three measurements (in yards). How many square yards of carpet will that be?

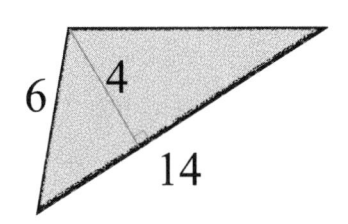

285. Outside her mansion she wanted a triangular concrete courtyard where she could have the gardener plant little flowers along the 30-meter perimeter.

How many square meters of concrete will have to be poured?

(A meter is a little longer than a yard.)

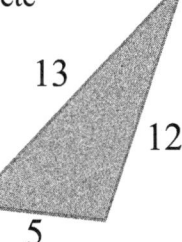

307. Darlene wanted to fly a giant flag over her mansion. When she told that to Joe, he suggested that the flag should have a fish on it. That would show that a fisherman (Joe) lived there. Darlene had other ideas.

The flag would be made out of steel. Doing it that way would mean that people could always read her flag even when the wind wasn't blowing.

Joe asked Darlene what the "𝔇" on the flag stood for. 𝔇arlene thought for a moment and lied, "It stands for Devoted wife."

Here are some of the dimensions of that steel flag (in feet). What is its area?

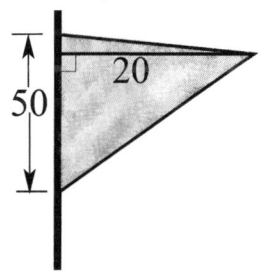

Chapter Twenty-five  Area of a Triangle

## Second part: the 𝔐ixed 𝔅ag: a variety of problems from this chapter and previous chapters

448. (continuing the previous problem) If the steel flag weighs 5 pounds per square foot, how much would the 500 square foot flag weigh?

649. (continuing the previous problem) Convert the weight of that flag into tons. (1 ton = 2,000 pounds)

766. Fill in the blanks.
 53.814 = 5_____ + 3_____ + 8_____ +1_____ + 4_____

798. Flag poles are designed to support cloth flags weighing several pounds, not steel flags weighing tons. You can guess what happened.

 The 𝔇 flag sliced through the roof of the mansion.

intelligent guess

 The million-dollar mansion became a $700,000 mansion.  Draw a bar graph.

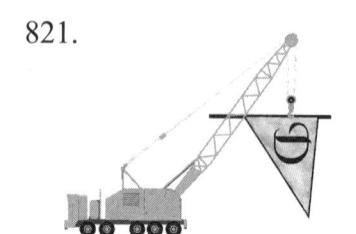

821.  They got a large crane to pull the 𝔇 flag out of the roof of the mansion.

 The crane set it on the ground and the workers built a circular fence around it so that people wouldn't accidentally walk into it and cut themselves on the sharp edges of the steel flag.

 The fence was 31 meters long.  What was the radius of that circle?  (Use 3.1 for π.)
(If you need a hint, see #102 in Chapter 14.)

# Chapter Twenty-six
# Area of a Parallelogram

## First part: Problems from this chapter

172. Find the area of this parallelogram.

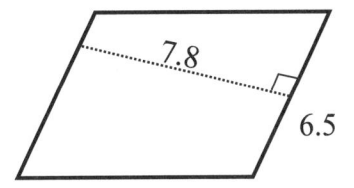

268. Joe measured this parallelogram. Is it possible to find the area just using his measurements?

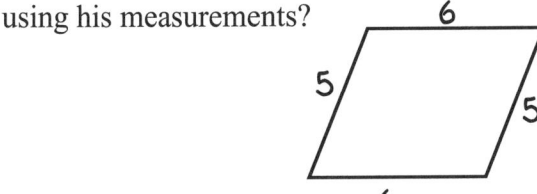

348. Darlene was designing her mansion that she would live in after she married Joe. She made one of the rooms in the shape of a parallelogram. She wanted the sides to be 7.4 meters and 10.5 meters and the length of the short diagonal to be 13.9 meters.

*Without doing the arithmetic* describe *how* you could find the area of this room.

(This is a hard question. Less than 10% of my readers will be able to figure it out.)

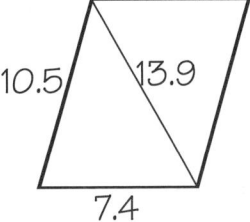

569. Darlene wanted a giant room for dancing. She had read in one of her bridal magazines that those rooms are called ballrooms.

Instead of reading the article, she decided to make one of the rooms in her mansion the shape of a ball. She didn't realize that all the chairs and tables and people in her ballroom would tumble down to the bottom of the ball. (In geometry balls are called spheres. The volume of a sphere = $(4/3)\pi r^3$.) What would be the volume of Darlene's ballroom if its radius were equal to 40 feet? (Use 3.1 for $\pi$.)

### Chapter Twenty-six  Area of a Parallelogram

## Second part: the 𝔐ixed 𝔅ag: a variety of problems from this chapter and previous chapters

146. Joe had always been fascinated by gumball machines. You put in a coin and twist the lever and out would pop a gumball. Joe had $97.50 saved up. He spent 40% of that on a gumball machine. How much did he spend?

200. Joe didn't especially like gumballs. They weren't sweet enough. He liked jelly beans. When he got his new gumball machine, he emptied out the gumballs. The glass sphere on his machine measured 10 inches across. (diameter)
  What was the volume that the machine could hold?
  $V_{sphere} = (4/3)\pi r^3$  (Use 3 for $\pi$.)

429. Joe wondered how many jelly beans he could stuff into his gumball machine. He knows the volume of the machine (from the previous problem). He does a little measuring and finds that 21 jelly beans occupy 6 cubic inches.
  This is a conversion factors problem. We want to convert 500 cubic inches into jelly beans.

523. Darlene often told Joe that he would have to make a lot more money than he was making right now. The only thing that Joe could think of is doing more fishing.
  Darlene pointed to Joe's new ~~gumball~~ jelly bean machine. She told him, "Fill it with jelly beans and stick it with the other vending machines in the hallway outside of Fred's office. You could make a pile of money."
  Tears started to form in Joe's eyes. "But . . . but . . . people will eat all my jelly beans!"
  "RELAX!!!" she screamed at him. "If jelly beans cost you a penny apiece and you sell them for a nickel, you will have a lot more money to buy many more jelly beans."
  Joe didn't understand everything Darlene was telling him, but he did what she said. There were now ten vending machines on the third floor of the Math Building, including Joe's jelly bean machine.
  If the starting set (the domain) is {Fred, Darlene, Joe, Betty, Alexander} and the rule is *Assign to each of these persons the number of jelly beans they purchased in the first 12 hours that Joe's machine was installed on the third floor.* Is this a function?

## Chapter Twenty-seven
## 13 Is What Percent of 52?

### First part: Problems from this chapter

If you know both sides of the *of* you multiply.   40% of 55 ➡ 0.4 × 55
If you don't know both sides of the *of*
    you divide the number closest to the *of*
        into the other number.   4 is what percent of 24?
            4 = ?% of 24   ➡   $24\overline{)4}$

176. There were 1,310 jelly beans in Joe's jelly bean machine when C.C. Coalback stole it. Before the theft Joe owned 65,500 jelly beans. What percent of Joe's jelly beans had been taken?

218. Joe's $39 jelly bean machine was 30% of the value of all the things Coalback took that night. How much had he taken?

323. The next morning Coalback was having breakfast and thinking about the things he had purloined (to purloin is to steal   purr-LOIN) that week. He didn't like jelly beans. They weren't sweet enough for him. He was having a bowl of sugar for breakfast.

      He had no use for a jelly bean machine. He knew that if he filled it with jelly beans and put it on, say, the second floor of the Art Building, that would be like *honest work*—something he'd never do. "And, besides," he thought, "someone might steal it."

      He decided to sell his $624 stash at a 35% discount to Sergeant Snow. How much would Snow pay him?

492. The next day was a fishing day for Joe. He wasn't going to worry about school or about his jelly bean vending business. He devoted one day out of seven to sitting in his boat. One is what percent of seven? Round your answer to the nearest percent.

593. So far in Coalback's life, he had spent 20 years in prison. He's now 36 years old. What percent of his life has he spent behind bars? Round your answer to the nearest percent.

Chapter Twenty-seven  13 Is What Percent of 52?

## Second part: the 𝔐ixed 𝔅ag: a variety of problems from this chapter and previous chapters

178. When Coalback was 21, he was sentenced to 2 years and 5 weeks in prison. After he had served 9 weeks, how long did he have left on his sentence? (Assume 1 year = 52 weeks.)

241. Let R = {eagle, canary, parrot}.
    Let S = {eagle, moose}.
    A) Is R ⊂ S true?
    B) R − S = ?
    C) R ∪ S = ?
    D) R ∩ S = ?

367. What is the largest element of this set? {x | x is a natural number and 44 < x < 66 and x is divisible by 3}

478. Express $\frac{5}{6}$ as a decimal. Put a bar over the repeating part.

530. The police are chasing Coalback down an alley. Coalback is running at the rate of 6.7 feet per second. (This is slow. Bowls of sugar for breakfast can really slow you down.)
    There is a door 53.6 feet from him. If he can get to that door, he can open it and hide. How long will it take him to reach that door? (Recall d = rt and $\frac{d}{r}$ = t)

607. Coalback never reaches the door. A policeman catches him. We know that $r_{coalback}$ = 6.7 feet per second.
    Which one of these must be true?
    A) $r_{cop} < r_{coalback}$
    B) $r_{cop} = r_{coalback}$
    C) $r_{cop} > r_{coalback}$

754. Convert $8\frac{3}{8}$ into a decimal.

846. (continuing the previous problem) If Coalback can run 6.7 feet per second and the policeman can run $8\frac{3}{8}$ feet per second, Coalback's speed is what percent of the policeman's speed?

# Chapter Twenty-eight
## Ratio

### First part: Problems from this chapter

The **ratio** of 7 mice to 3 cats can be written as 7 mice:3 cats or $\frac{7 \text{ mice}}{3 \text{ cats}}$ or 7:3 or $\frac{7}{3}$

147. If the ratio of apples to cherries is 2:11, then what is the ratio of cherries to apples?

272. In Chapter 3 of *Life of Fred: Beginning Algebra Expanded Edition* we will meet **continued ratios**. They are even easier than regular ratios. When Coalback was 19, he committed 4 robberies, 3 car thefts, and 14 drug deals. The continued ratio is 4:3:14. Continued ratios are easier than regular ratios because they can only be written with colons (: ← This is a colon) and not as fractions.

   Darlene has 48 bottles of red nail polish, 6 of blue polish, and 2 of yellow polish. Express this as a continued ratio.

385. The ratio of sunny days to cloudy days in Kansas is $\frac{7}{8}$. Express this ratio using colons.

413. In Fred's geometry class he once noticed that 24 students were wearing white socks and 36 students were wearing blue socks. He said that the ratio of white-sock wearers to blue-sock wearers was 2:3.
   Explain how he got that answer.

Chapter Twenty-eight  Ratio

# Second part: the 𝔐ixed 𝔅ag: a variety of problems from this chapter and previous chapters

167. What is the smallest element of {x | x is a prime number and x > 13}?

313. Goldbach's first conjecture was that every even number greater than 2 can be written as the sum of two primes. Express 46 as the sum of two primes.

405. When Darlene was planning her mansion that she would have after she married Joe, she thought it would be nice to have a little area for Joe to exercise in. It would be important for him to keep fit so that he could spend long hours at work to pay for her mansion.
    She drew the plans. The gym would be 35 yards by 80 yards. What is the area of this rectangle?

542. Hardwood floors for a gym are expensive. They can cost $70 per square yard. How much will the flooring cost for Darlene's gym?

688. Off in a corner of the gym Darlene would have the workers paint a circle. It would be 40 inches across. That would be a perfect spot for Darlene to play marbles with her girl friends. What would be the area of that circle? (Use 3.1 for $\pi$.)

734. Change $4.35 into cents.
    Change 4.35 into a percent.

872. Darlene knew exactly how Joe would spend his time after they got married. Each day he would spend . . .
            12 hours at work
             1 hour at the gym
             8 hours sleeping
             3 hours cleaning the mansion
    Draw a pie chart (circle graph).

# Chapter Twenty-nine
## Ordered Pairs

### First part: Problems from this chapter

Functions can be written as ordered pairs. If the domain (the starting set) is {Darlene, Joe, Fred, Coalback} and the rule is *Assign to each person the number of lies they have told in the last month*, then this function could be written as

    Darlene → 1
    Joe → 0
    Fred → 0
    Coalback → 5,774

or it could be written as ordered pairs: (Darlene, 1), (Joe, 0), (Fred, 0), (Coalback, 5774).

Sometimes it is written as a *set of ordered pairs*: {(Darlene, 1), (Joe, 0), (Fred, 0), (Coalback, 5774)}.

156. Is this a function?
    (✿, 8), (✉, 8), (✭, 30), (✉, 5)

206. This is a function. {(a, rat), (b, house), (c, moon)}. What is the domain of this function?

329. Which of these are functions?
    A) {(hat, rain), (shoe, rain), (glove, rain)}
    B) {(7, grape), (7, peach)}
    C) {(1, 2), (2, 3), (3, 4), (4, 5)}
    D) {(Newton, 77), (Archimedes, 77)}

624. The starting set is {m, n, p}.
    m → blue
    red is the image of n
    p is mapped to white

Write this function as a set of ordered pairs.

Chapter Twenty-nine  Ordered Pairs

**Second part: the 𝔐ixed 𝔅ag: a variety of problems from this chapter and previous chapters**

179. Darlene read an article in BARBEQUED BRIDE magazine.

### Get Him to Cook!
If you buy a BBQ grill, you can get him to do all the cooking. Tell him that working on a BBQ is a he-man thing to do.

Don't tell him that it's just cooking except that
  it's outside,
  it's dirtier, and
  it's easier to get burned.

Four thousand women read that article, and 27% of them bought a BBQ grill that week. How many was that?

194. Darlene, of course, bought a grill. She was eager to teach Joe how to BBQ. There was an ad in the KITTEN Caboodle newspaper: Beef Ribs! Normally $2 a pound. Now 35% off. How much is the sale price?

214. The trap was set. Darlene phoned Joe and told him, "Hello, Joe. You are my he-man. Are you hungry?" She knew what the answer was. That's called a rhetorical (re-TORE-ee-cull) question—a question that needs no answer.

She continued, "I've just bought ten pounds of beef ribs."
10 × $1.30 = $13.00.

Darlene could tell that Joe's mouth was hanging open.

Joe was in his boat, 300 feet from shore. He was heading toward shore at the rate of 5 feet per second.

Once he got to shore, he hopped on his bicycle and peddled at the rate of 25 mph. It was 5 miles from the shore to Darlene's apartment.

How long before Joe was knocking on her door?

## Chapter Twenty-nine  Ordered Pairs

342. "When do we eat?" were the first words out of Joe's mouth.

Darlene asked, "I just bought these ribs. I think they'll taste better if we cook them." Darlene had said *we*, but she was thinking *you*.

She had already turned on the grill. It was ready for cooking. She turned to Joe and asked him to put the meat on the grill. She was going to teach him how to do man-cooking. She was sure that he had never done it before. The only cooking that Joe knew how to do was opening bags of jelly beans.

"Be sure to take the ribs out of the package before you put them on the grill" she instructed.

Joe piled the ribs on the grill and waited. After two minutes he asked, "Are they done?"

Darlene knew that they wouldn't be done in two minutes. She asked her cell phone, and it told her: `Beef ribs take 500% longer than two minutes to cook on a grill.`

What is the total time to cook ribs?

641. Joe took the burned ribs off the grill. He used the tongs after he realized that the ribs were too HOT to grab with his hands. He wrapped them in a cloth that he found on the table and threw them in the garbage.

That "cloth" was Darlene's new silk scarf.

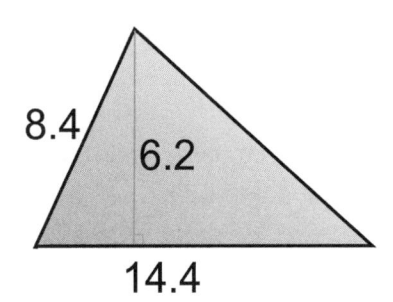

What was the area of her scarf?

# Chapter Thirty
# Graphing

## First part: Problems from this chapter

253. This is a function: {(ball, string), (mouse, 5), (door, Germany)}. What is the set of all first coordinates called?

358. Graph {(3,1), (2, 2), (1, 4)}.

423. Convert this graph into a set of ordered pairs.

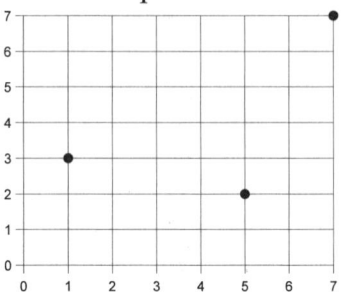

595. Graph {(1, 200), (2, 600), (3, 100)}. You probably won't use the same scale for the horizontal x-axis and the vertical y-axis.

694. When you graph some ordered pairs, you have to draw in the horizontal x-axis and the vertical y-axis.

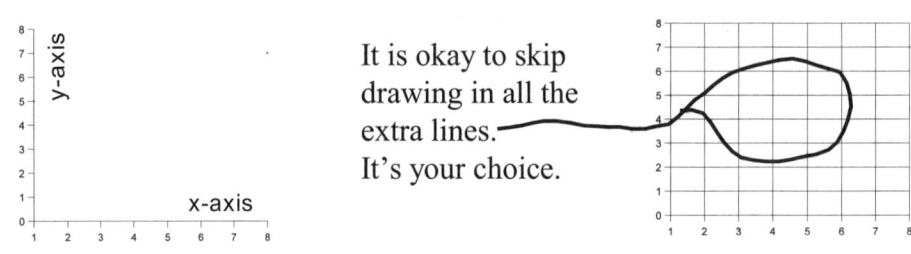

It is okay to skip drawing in all the extra lines. It's your choice.

Is it okay to skip putting in the numbers (the scale) on the x-axis and the y-axis?

### Chapter Thirty  Graphing

## Second part: the 𝔐ixed 𝔅ag: a variety of problems from this chapter and previous chapters

231. Darlene told Joe that she was going to have a birthday party for herself. Joe scratched his head and asked, "I thought your birthday was several months from now."

Darlene told him, "This will be just for practice."

Joe didn't understand. "Practice for what?"

She smiled. "This is practice for your giving me gifts. To start with, please blow up this birthday balloon for me."

Joe began to blow. He noticed that it had her name on it, but he didn't say anything about that. Instead, he asked how big he should make the balloon.

"Just keep blowing. I'll tell you when to stop."

The phone rang and Darlene got into a conversation with one of her girlfriends. Joe kept blowing. When the radius of that balloon ball was four feet, how much air was in it? ($V_{sphere} = (4/3)\pi r^3$. Use 3 for $\pi$.)

a balloon that can't be popped

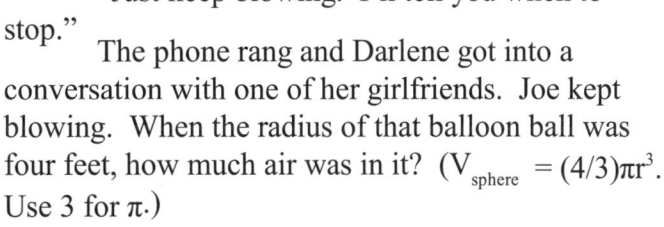

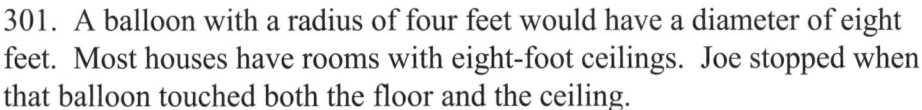

301. A balloon with a radius of four feet would have a diameter of eight feet. Most houses have rooms with eight-foot ceilings. Joe stopped when that balloon touched both the floor and the ceiling.

A full blow by Joe was 2 quarts of air.

The balloon was 256 cubic feet.

How many blows did Joe do in order to inflate that balloon? Round your answer to the nearest blow.

(A dry quart is equal to about 67 cubic inches. There are $12^3$ cubic inches in a cubic foot. Use conversion factors to convert one balloon into blows by Joe.)

> *Intermission*
>
> In the Imperial (English) system, a dry quart is equal to about 67 cubic inches.
>
> A liquid quart is equal to about 58 cubic inches.
>
> So a quart of milk has less volume than a quart of wheat.
>
> This is just another reason why the metric system keeps people from going nuts.

a quart?

## Chapter Thirty  Graphing

442. Joe asked, "I'm supposed to give you gifts for your birthday party?"

Darlene said, "Of course. That's why people have birthday parties—to get gifts.*"

[That's a footnote!]

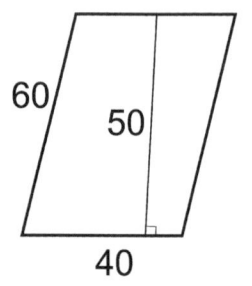

"Okay," Joe said. "I have designed this wonderful tennis court for you. I have even put in the dimensions so that the workers can build it for your mansion. The numbers are all in feet."

"That thing is crooked! How can I play tennis with my girlfriends when it isn't a rectangle?"

Joe explained, "I drew it that way because I'm smart. If there is a wind blowing, then the ball will stay in the court."

What is the area of that parallelogram?

---

* Is that really true? Not everything you read in a book is literally true. If you think it is, then believe this: $2 + 2 = 5$.

Darlene and Joe, if you haven't figured it out already, do not understand a lot of things. Darlene thinks that they will be able to afford a mansion after she marries Joe, but neither of them have the education nor the skills necessary to earn a lot of money.

The thought that people give birthday parties just to get a lot of gifts is something that a two-year-old might believe.

In good high school English classes, you will learn about **irony**. When authors are writing the opposite of what is obviously true, they are using irony.

Years ago Jonathan Swift wrote a little essay entitled "A Modest Proposal." I read that essay in high school. In the early part of his essay, he wrote about how difficult life was for the poor starving people of Ireland. Then he suggests that they sell their one-year-olds as "a most delicious nourishing and wholesome food, whether stewed, roasted, baked, or boiled. . . ." This is total irony. He didn't mean it! For high school students who have learned about irony, Swift's essay is a total giggle. However, I would never let little kids read his essay. They might think he was being serious.

# Chapter Thirty-one
# Nine Conversions

## First part: Problems from this chapter

237. Convert $\frac{3}{8}$ into a percent.

319. Change $\frac{5}{6}$ into a percent.

❈ ❈ ❈

Here is how you change 66⅔% back into a fraction.

$66\frac{2}{3}\% = \frac{66\frac{2}{3}}{100}$     Dividing by a 100 is the same as moving the decimal two places to the left.

$= 66\frac{2}{3} \div 100$

$= \frac{200}{3} \div 100$     When you divide (or multiply) mixed numbers, you have to first change them into improper fractions. Changing 66⅔ into an improper fraction means "3 times 66 ... plus 2"

$= \frac{200}{3} \div \frac{100}{1}$

$= \frac{200}{3} \times \frac{1}{100}$

$= \frac{2}{3}$

❈ ❈ ❈

461. Now it's your turn. Change 16⅔% into a fraction.

## Chapter Thirty-one  Nine Conversions

# Second part: the 𝔐ixed 𝔅ag: a variety of problems from this chapter and previous chapters

263. Joe was looking at the super mansion that Darlene was designing for herself. He had no idea where the money would be coming from. He let Darlene do that kind of thinking. He showed her a picture of the fishing boat of his dreams. (= $80,000)

Joe's dream boat

Darlene had other plans for Joe. After they had purchased the mansion, a new electric car for Darlene, and a new wardrobe for her, she estimated that they might have $10,000 left over for Joe to spend on their honeymoon and maybe a canoe.
$10,000 is what percent of $80,000?

381. That canoe might be worth $300. (It's not new.) Joe's original dream boat is worth $80,000. What is the ratio of the value of his dream boat to that canoe?

466. {(boat, dish), (dish, soap), (sky, rope)} is a function.
What is the image of dish under this function?

515. {(boat, dish), (dish, soap), (sky, rope)} is a function.
Is the set of second coordinates {dish, soap, rope} the domain of that function?

668. Is this the graph of a function?

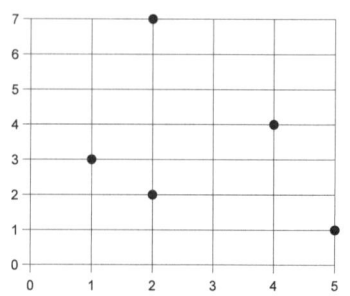

965. Change 87½% into a fraction.

76

# Chapter Thirty-two
# Elapsed Time

## First part: Problems from this chapter

239. **Ralph's Jelly Beans** don't taste as good as other brands, but they are cheap and they come in the popular Snack Size.
   When Joe got to Fred's arithmetic class, he had already eaten three-fourths of a bag. While Fred lectured, he finished that bag and ate 2⅓ pounds in a second bag. How much did Joe eat during Fred's lecture?

259. **Ralph's Perfume** doesn't smell very good. (In fact, it stinks.) But it is cheap and easy to apply. On Tuesday Darlene had used 70% of a bottle. Today, Wednesday, she finished that bottle and used three-eights of a second bottle.
   What percent of a quart had she used today?

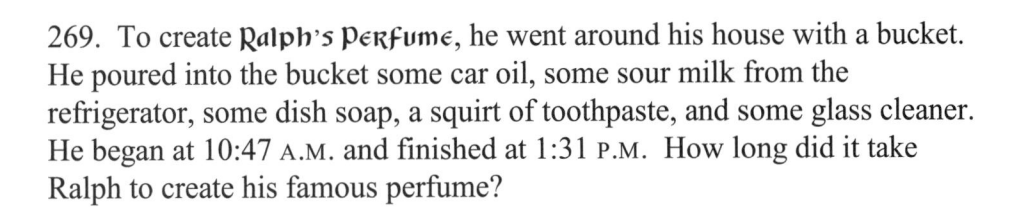

269. To create **Ralph's Perfume**, he went around his house with a bucket. He poured into the bucket some car oil, some sour milk from the refrigerator, some dish soap, a squirt of toothpaste, and some glass cleaner. He began at 10:47 A.M. and finished at 1:31 P.M. How long did it take Ralph to create his famous perfume?

354. **Ralph's Perfume** was also good for clearing clogged drains. If your sink drain is full of water, just pour an ounce of **Ralph's** into the water and your drain will open in 45 minutes.
   Joe's drain had been clogged up with jelly bean gunk.
   At 3:52 P.M. he added an ounce of **Ralph's** and waited. When should Joe expect the drain to open?

Chapter Thirty-two  Elapsed Time

## Second part: the 𝔐ixed 𝔅ag: a variety of problems from this chapter and previous chapters

273. Graph {(2, 1), (3, 12), (4, 2)}.

316. 42.6 − 3.008 = ?

335. Darlene read in BRIDE'S GUIDE TO POLAND magazine about the wonderful music you can hear when you visit that country.

"We gotta go there after we get married," Darlene told Joe.

"Married? Married to who*?" Joe asked.

Darlene couldn't believe those questions. After all the years that Darlene had been working on Joe, he still didn't realize Darlene's intentions. She figured, correctly, that Joe hadn't been paying attention.

Joe popped a couple of jelly beans into his mouth, and Darlene explained s l o w l y, "We might get married to each other. You and me. Like a couple. Then we could go on our honeymoon to Poland. Would you like that?"

Joe chewed and asked, "Like what?"

"What I've been talking about!" Darlene was going out of her mind.

Joe shrugged his shoulders. "I don't know that much about Polish music. Couldn't we just buy a CD and listen to it?"

If Darlene had spent 3 years so far working on getting Joe to propose marriage and if she was 6% of the way toward that goal, how long would the whole process take?

357. Joe couldn't figure out why Darlene was crying. To make her feel better he went to the store and bought the compact disc "Polish Polkas." He would play that while he was out fishing alone. That would make her feel better. It cost him 90% of the money he had in his pocket ($20). How much did that CD cost?

---

* It should be, "Married to whom?"

# Chapter Thirty-three
# Probability

## First part: Problems from this chapter

368. Joe had a handful of jelly beans. There were 15 blue ones, 5 black ones, 5 orange ones, and 75 green ones. Without looking, he popped one of them into his mouth. What is the probability that it was orange?

384. (continuing the previous problem) What is the probability that the jelly bean will be either black or green?

403. (continuing the previous problem) What is the probability that the jelly bean won't be either black or green?

417. Darlene has a lot of nail polish bottles. If she picks one at random, there is a 7% chance that it will be blue. What is the chance that it won't be blue?

430. If there is a 100% probability that Joe won't propose to Darlene this week, what is the probability that he will propose?

455. When two things are entirely unrelated—such as the color of nail polish that Darlene chooses on a particular day and the number of crimes C.C. Coalback commits on a particular day—and you want to find the probability of BOTH happening, you multiply.

For example, suppose there is a third of a chance that Darlene will wear red nail polish and two-fifths of a chance that Coalback will commit more than a dozen crimes.

Then the probability that Darlene will wear red nail polish *and* Coalback will commit more than a dozen crimes is $\frac{1}{3} \times \frac{2}{5} = \frac{2}{15}$

The probability that the next fish Joe catches is a tuna is $\frac{3}{80}$

The probability that Fred will own a car sometime in the next ten years is $\frac{20}{21}$

What is the probability of both of these things happening?

Chapter Thirty-three  Probability

## Second part: the 𝔐ixed 𝔅ag: a variety of problems from this chapter and previous chapters

462. There is a 3% chance that Joe will sell his boat and give up fishing. There is a 7% chance that Fred will win the Fields Metal this next year. What is the probability that both will happen?

(The hard part of this question is: How do you multiply two percents together? We have never done that before.)

471. Darlene was reading PIANOS FOR THE ACTIVE BRIDE magazine. Almost all the pianos in that magazine were the standard 88 keys.

Then she found the perfect piano for her future mansion. It had 50% more keys than a standard piano. How many keys did it have?

511. Darlene took Joe to a piano store. The salesman took Joe into the bathroom and made him wash the fish oil off his hands before he touched any of the pianos.

While Darlene and the salesman were talking, Joe went to one of the regular (88-key) pianos and started to hit each of the keys. He could hit 11 keys in 15 seconds. How long would it take him to hit all 88 keys?

532. After washing his hands, but before playing the piano, Joe had a snack of purple jelly beans. Thirty-two of the 88 keys now had purple marks on them. What percent of the keys were marked by Joe? Round your answer to the nearest percent.

570. At 9:35 A.M. they entered the piano store. At 9:58 A.M. Joe fell asleep under one of the pianos. At 2:30 P.M. Joe woke up (and hit his head as he stood up). At 2:51 P.M. Darlene finished talking with the salesman and they left. How long had Joe slept?

672. Create a function with domain = {Joe, Darlene} and with answers in the set {egg, rice}. (There are four possible answers to this question.)

# Complete Solutions and Answers | 100–107

100. Find the first whole number greater than a million that is divisible by both 3 and 5.

  The numbers greater than a million that are divisible by 5 are: 1,000,005, 1,000,010, 1,000,015, 1,000,020, 1,000,025. . . .

  The first one on that list—1,000,005—is divisible by 3 because the sum of its digits is divisible by 3.

102. The wall would be circle with a length of 22.32 miles. What would be the diameter?

$$3.1 \overline{)22.32}$$

$$\begin{array}{r} 7.2 \\ 31 \overline{)223.2} \\ \underline{217} \\ 62 \\ \underline{62} \end{array}$$
  The diameter of the wall would be 7.2 miles.

104. Express 466 as the sum of two primes.

  Looking on that list of primes, we see that 463 is a prime.
  466 = 463 + 3  There might be other answers.

105. Fred is three feet tall and Alexander is six feet tall.

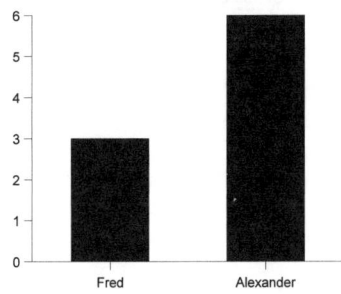

107. None of these numbers are prime numbers.
  400002, 9067097930, 555455554, 7751258, 33388878878878
  What is an easy way to show that they are all composite?

  All of these numbers are even. (Their last digits are either 0, 2, 4, 6, or 8.) The only even prime number is 2.

| 108–123 | **Complete Solutions and Answers**

108. Fill in the blanks.

85.79 = _8_ tens + _5_ ones + _7_ tenths + _9_ hundredths

110. Suppose the starting set is {Fred, Alexander, Betty, Joe, Darlene}. Would the rule *Assign each person to the woman who gave birth to them* be a function?

    Yes. Fred had exactly one birth mother. Alexander had exactly one birth mother. etc.

114. The rule *Put the 5-gram jelly beans in one pile, the green ones in a second pile, and all the rest in a third pile* is not a function. Why?

    Each member of the domain must be assigned to exactly one answer (= one pile). Which pile would a 5-gram, green jelly bean be assigned?

117. 5.7 − 1.27 = ?

$$\begin{array}{r} 5.7 \\ -1.27 \\ \hline \end{array} \quad \Rightarrow \quad \begin{array}{r} 5.70 \\ -1.27 \\ \hline 4.43 \end{array}$$

5.7 means the same thing as 5.70.

119. Fill in the blanks.

602.1 = 6 _hundreds_ + 0 _tens_ + 2 _ones_ + 1 _tenth_

121. Change $\frac{5}{8}$ into a decimal. (That's done by dividing 5.00000 by 8.)

$$\begin{array}{r} 0.625 \\ 8\overline{)5.00000} \\ \underline{48} \\ 20 \\ \underline{16} \\ 40 \\ \underline{40} \\ 0 \end{array}$$

$\frac{5}{8} = 0.625$

123. Multiply each of these by ten.

        3.45   9.9   0.002   550

        34.5   99.   0.02   5500

# Complete Solutions and Answers | 125–136

**125.** By noon at day she had walked 290 feet. How much farther will she need to walk in order to walk 3 miles? (1 mile = 5,280 feet)

```
   3 miles                  2 miles   5280 feet
-         290 feet       -           290 feet
                           2 miles   4990 feet
```

Darlene has 2 miles 4,990 feet yet to walk.

**126.** If it would take Wolfie 0.3 hours, how many minutes is that?

$$\frac{0.3 \text{ hours}}{1} \times \frac{60 \text{ minutes}}{1 \text{ hour}} = \frac{0.3 \text{ hours}}{1} \times \frac{60 \text{ minutes}}{1 \text{ hour}} = 18 \text{ minutes}$$

**128.** Change 0.25 into a fraction.

$$0.25 = \frac{25}{100} = \frac{1}{4}$$

**130.** $6.4 + 9.7 = ?$

```
   6.4
+  9.7
  16.1
```

**132.** Fill in the blanks as we change base 10 into base 2.

$$28_{\text{base 10}} = \underline{\ ?\ } \times 2^4 + \underline{\ ?\ } \times 2^3 + \underline{\ ?\ } \times 2^2 + \underline{\ ?\ } \times 2 + \underline{\ ?\ } \times 1$$
$$\phantom{28_{\text{base 10}} = }\ \ 16\ \ \ \ \ \ \ \ \ \ \ \ \ 8\ \ \ \ \ \ \ \ \ \ \ \ \ 4\ \ \ \ \ \ \ \ \ \ \ \ \ 2\ \ \ \ \ \ \ \ \ \ \ \ \ 1$$

$28 = 1 \times 16 + 1 \times 8 + 1 \times 4 + 0 \times 2 + 0 \times 1$

so $28_{\text{base 10}} = 11100_{\text{base 2}}$

**134.** Express 204 as the sum of three primes.

199 is a prime. That leaves (204 – 199 = 5) five more.
204 = 199 + 3 + 2

**136.** The president of KITTENS University makes 70% of the total salaries and the teachers make 25% of the total salaries and the secretaries make 5% of the total salaries.

Salaries at KITTENS

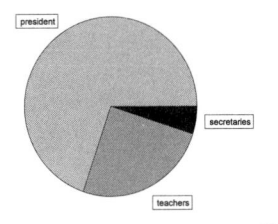

83

## 137–139 Complete Solutions and Answers

**137.** Darlene saw an ad. It was a square plot of land.

A) How many square miles is that?

We want to convert 10,240 acres into square miles. Conversion factor!

$$\frac{10240 \text{ acres}}{1} \times \frac{1 \text{ square mile}}{640 \text{ acres}} = 16 \text{ square miles}$$

B) What is the length of the side of that square?

Each side must be 4 miles long. (Arithmetic: 4 miles × 4 miles = 16 square miles)

**139.** The diameter of the plate is 12". What is its area?

If the diameter is 12", then the radius is 6".
$A = \pi r^2 = \pi \times 6 \times 6 = 36\pi$ square inches.

$36\pi$ square inches is the *exact answer*. That's the answer that mathematicians are happy with.

Despite rumors to the contrary, mathematicians are humans. Like most humans, they don't enjoy multiplying numbers. They are happy knowing that $\pi \times \pi = \pi^2$. None of them (that I know of) will approximate $\pi$ with 3.14159265 and then for fun compute $\pi^2$ on a napkin in a pizza restaurant . . .

```
            3.14159265        eight digits to the right of the decimal
         × 3.14159265         eight digits to the right of the decimal
           1570796325
           1884955590
            628318530
           2827433385
           1570796325
            314159265
           1256637060
            314159265
            942477795
      9.8696043785340225  ⟹   9.8696043785340225
                              sixteen digits to the right of the decimal
```

# Complete Solutions and Answers  | 140–141

140. Enough seed to plant a circle with an area of 75 square miles. What would be the radius of that circle?

We know that the area of a circle = $\pi r^2$.

$75 = \pi r^2$

We will use 3 for $\pi$.   $75 = 3r^2$

Since you don't know algebra yet, the only thing you can do is try out some numbers and see if they work.

If we let r be equal to 3, we get $75 = 3\times3\times3$    $75 \stackrel{?}{=} 27$  No.
If we let r be equal to 4, we get $75 = 3\times4\times4$    $75 \stackrel{?}{=} 48$  No.
If we let r be equal to 5, we get $75 = 3\times5\times5$    $75 \stackrel{?}{=} 75$  Yes!

Once we get to algebra, we don't have to guess.

We start with                          $75 = 3r^2$
Divide both sides by 3                 $25 = r^2$
Take the square root of both sides     $5 = r$

❀   ❀   ❀

In a million years you couldn't solve  $72 = 3r^2$  by guessing, but algebra will work.

We start with                          $72 = 3r^2$
We divide both sides by 3              $24 = r^2$
Take the square root of both sides     $\sqrt{24} = r$

And you ask, "What is $\sqrt{24}$ equal to?"

On a hand calculator, you type in ⌶ ५ and hit the √‾ key.

Out will pop ५.8989794, which is more accuracy than you would need in everyday life.

I have a big calculator. It says $\sqrt{24}$ = ५.8989794855663561963945681494118

141. How many square yards of carpet will that be?

The base is 14 yards. The height is 4 yards.
A = (½)bh = ½ × 14 × 4 = 28 square yards.

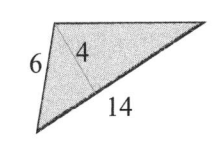

85

| 142–146 | Complete Solutions and Answers

142. In her first month of collecting, when she was just starting out, she would buy 400 vases. In the second month she would have learned how to really buy vases and would buy 34% more than she did in the first month. How many vases did she imagine buying in the second month?

**The hard way:**

First, find out how many more vases she would buy.

34% of 400 = 0.34 × 400 = 136

Second, add that to the first month's 400

136 + 400 = 536

**The easier way:**

A 34% gain means the original 400 (100%) plus an extra 34% 100% + 34% = 134%

134% of 400 = 1.34 × 400 = 536

In the second month, Darlene imagined buying 536 vases.

144. Let T = {x | x is a vase and x weighs less than three pounds}
Let F = {x | x is a vase and x weighs less than five pounds}

A) Is T ⊂ F true?

Yes. Every vase that weighs less than three pounds must weigh less than five pounds.

B) T ∪ F = F

The set of all vases that either weigh less than three pounds or weigh less than five pounds is the set of all vases that weigh less than five pounds.

C) T ∩ F = T

The set of all vases that both weigh less than three pounds and weigh less than five pounds is those that weigh less than three pounds.

D) T – F = ∅, which can also be written as T – F = { }.

Start with the set of all vases weighing less than three pounds and eliminate from that set all vases weighing less than five pounds and you have nothing left.

146. 40% of $97.50 = 0.4 × 97.50 = 39. He spent $39.

# Complete Solutions and Answers | 147–155

147. If the ratio of apples to cherries is 2:11, then what is the ratio of cherries to apples?

It is 11:2. It would also be correct to write 11 cherries:2 apples or $\frac{11}{2}$ or $\frac{11 \text{ apples}}{2 \text{ cherries}}$

150. At the King KITTENS mall, Darlene bought some nail polish ($17.77), some lipstick ($34.99), and some hair spray ($15.95).

Joe bought some fishing line ($16.55), some bait ($8), and some bags of jelly beans ($43.45).

Who spent more money?

```
Darlene spent  17.77          Joe spent  16.55
               34.99                      8.00
             + 15.95                   + 43.45
               68.71                     68.00
```

Since $68.71 > $68.00, Darlene spent more.

153. It was normally $80,000, but it was on sale for 20% off of that price.

**The hard way:**

First, find out how much you save.

20% of $80,000 = 0.2 × 80,000 = $16,000.

Second, subtract the savings to get the sale price.

80,000 − 16,000 = $64,000.

**The easier way:**

If you are saving 20%, you will be spending 80%.

80% of 80,000 = 0.8 × 80,000 = $64,000.

155. If the circumference is about three times as long as the diameter, what would be the circumference of a circle whose radius was 5?

If the radius is 5, then the diameter would be 10.
If the diameter is 10, then the circumference would be about 30.

| 156–160 | Complete Solutions and Answers

156. Is this a function?

(✿, 8), (✉, 8), (★, 30), (✉, 5)

No. ✉ has been assigned two different answers. The definition of a function is any rule that assigned *exactly one answer* to each member of the starting set (the domain).

157. Its radius will be 0.8 miles. Find the area. Use 3.1 for $\pi$. Round your answer to the nearest tenth of a square mile.

$A = \pi r^2 \approx 3.1 \times 0.8 \times 0.8 = 1.984$ square miles

$1.984 \doteq 2.0$ square miles

A little note. When we measured length—such as the length of the radius—we use miles.

When we measure area we use square miles.

Of course, if we measured the radius in inches, the area would be in square inches.

If we measured the radius in meters (using the metric system), the area would be in square meters.

158. When chased, a dancing pizza can run 40 mph (miles per hour). How far can it run in 3.2 hours?

$d = rt = 40$ mph $\times 3.2$ hours $= 128$ miles

160. Express $87_{\text{base 10}}$ in base 2 (the binary number system).

$87_{\text{base 10}} = 1010111_{\text{base 2}}$

Here is the arithmetic:

$\phantom{+}1 \times 2\times2\times2\times2\times2\times2$
$+ 0 \times 2\times2\times2\times2\times2$
$+ 1 \times 2\times2\times2\times2$
$+ 0 \times 2\times2\times2$
$+ 1 \times 2\times2$
$+ 1 \times 2$
$+ 1 \times 1$

$1 \times 64 + 0 \times 32 + 1 \times 16 + 0 \times 8 + 1 \times 4 + 1 \times 2 + 1 \times 1$

# Complete Solutions and Answers  162–168

162. (continuing problem #102) If her mansion were at the center of the circle, how far would she have to walk to get from her mansion to the brick wall?

This would be the radius of the circle. A radius is half of the diameter. 7.2 ÷ 2 = 3.6 miles

164. Which of these are true?
  A) 5 ∈ {4, 5, 8}   True. 5 is a member of {4, 5, 8}
  B) red ∉ {blue, green}   True. red is not a member of (blue, green}
  C) cow is a member of {cow, horse, star}
      True. cow ∈ {cow, horse, star}
  D) duck is not an element of {✿, ✷, ✻}
      True. duck ∉ {✿, ✷, ✻}

166. Fill in the blanks.
    6.107 = __0__ tens + __6__ ones + __1__ tenth + __0__ hundredths + __7__ thousandths

    or you could have written:  ___ tens + __6__ ones + __1__ tenth + __0__ hundredths + __7__ thousandths  (leaving the tens place blank)

167. What is the smallest element of {x | x is a prime number and x > 13}?
    The prime numbers are 2, 3, 5, 7, 11, 13, 17, 19, 23, 29, 31, 37, 41, 43, 47, 53, 59, and so on. The smallest number in this set that is greater than 13 is 17.

168. Walking to class Joe ate 6 cherry jelly beans, 2 lime jelly beans, and 3 chocolate jelly beans. Draw a bar graph to illustrate this.

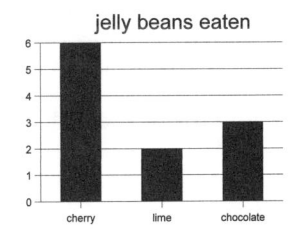

or you could draw it without the horizontal lines →

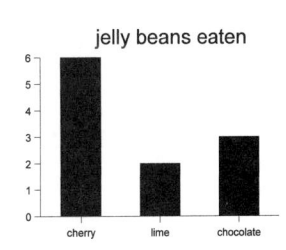

89

## 170–178 Complete Solutions and Answers

170. C.C. Coalback originally started out with $23.07. He spent $4.88 investing in all those funds. How much did Coalback have at this point?

$$\begin{array}{r} 23.07 \\ -\phantom{0}4.88 \\ \hline 18.19 \end{array}$$
He had $18.19 at this point.

172. A = bh = 6.5 × 7.8 = 50.7

174. KITTENS University pays a million dollars a year in salaries. The president makes 70% of the total salaries paid. How much does he get each year?

70% of 1,000,000 = 0.7 × 1,000,000 = 700,000.

*To multiply by a million you move the decimal 6 places to the right.*

The president makes $700,000 each year.

176. There were 1,310 jelly beans in Joe's jelly bean machine when C.C. Coalback stole it. Before the theft Joe had 65,500 jelly beans. What percent of Joe's jelly beans had been taken?

1,310 is what percent of 65,500?
65,500 is closest to the *of*.

$$\begin{array}{r} 0.02\phantom{00} \\ 65500\overline{)1310.00} \\ -\phantom{0}131000 \\ \hline 0 \end{array}$$

1,310 is 2% of 65,500

178. When Coalback was 21, he was sentenced to 2 years and 5 weeks in prison. After he had served 9 weeks, how long did he have left on his sentence? (Assume 1 year = 52 weeks.)

$$\begin{array}{r} 2 \text{ years } \phantom{0}5 \text{ weeks} \\ -\phantom{2 \text{ years }00}9 \text{ weeks} \\ \hline \end{array} \qquad 1 \text{ year } 52+5 \text{ weeks} \qquad \begin{array}{r} 1 \text{ year } 57 \text{ weeks} \\ -\phantom{1 \text{ year }00}9 \text{ weeks} \\ \hline 1 \text{ year } 48 \text{ weeks} \end{array}$$

After serving 9 weeks, Coalback had 1 year and 48 weeks left.

## Complete Solutions and Answers  179–191

**179.** Four thousand women read that article, and 27% of them bought a BBQ grill that week. How many was that?

$\qquad$ 27% of 4,000 = 0.27 × 4000 = 1,080 women bought a BBQ grill.

**180.** 7.8 × 9.7

```
        7.8        one digit to the right of the decimal
      × 9.7        one digit to the right of the decimal
      ─────
        546
        702
      ─────
       7566  ➡  75.66    two digits to the right of the decimal
```

**184.** Express $\frac{1}{6}$ as an unending decimal.

```
      0.16666666666666666666666 ...
   6) 1.0000000000000000000000000
      6
      ──
       40
       36
       ──
        40
        36
        ──
        etc.
```

$\frac{1}{6} = 0.16666666666666666666666 \ldots$

**186.** Change into cents.

$\qquad$ \$0.25 = 25¢ $\qquad$ \$3.95 = 395¢ $\qquad$ \$0.06 = 6¢

**191.** Darlene figured that those 17 friends would be happy to each pay their fair share of the \$8,029.61 wedding costs. How much would each owe?

```
         472.33
     17) 8029.61
         68
         ───
         122
         119
         ───
          39
          34
          ──
           56
           51
           ──
            51
            51
            ──
             0
```

Each friend would owe \$472.33.

> Darlene decided to "bill" each of her 17 friends after the wedding. By her doing that, none of them would be tempted to give cheaper wedding gifts.

91

| 193–200 | **Complete Solutions and Answers** |

193. At the Jelly Bean Rally Joe heard about a new flavor of jelly beans. They were called Banana Jelly Beans in a Box. A box cost $1.04. Joe paid $78. How many boxes did he get?

We want to convert $78 into boxes of banana jelly beans. How about a conversion factor?

$$\frac{\$78}{1} \times \frac{1 \text{ box}}{\$1.04} = \frac{\$78}{1} \times \frac{1 \text{ box}}{\$1.04} = 1.04\overline{)78.}$$

$$\begin{array}{r} 75. \\ 104\overline{)7800.} \\ \underline{728}\phantom{0} \\ 520 \\ \underline{520} \end{array}$$

Joe purchased 75 boxes of banana jelly beans.

194. Darlene, of course, bought a grill. She was eager to have Joe learn how to BBQ. There was an ad in the KITTEN Caboodle newspaper: Beef Ribs! Normally $2 a pound. Now 35% off. How much is the sale price?

35% off means that you pay 65% of the normal price.
65% of $2 = 0.65 × 2 = 1.3
The sale price is $1.30.

199. She walked around the edge of the table and estimated that it had a circumference of 47.1 feet.

If she and Joe sat across the table from each other, how far apart would they be?

The circumference C = 47.1. To get the diameter you divide by π.

$3\overline{)47.1}$ = 15.7     Joe and Darlene would be about 15.7 feet apart.

200. The glass sphere on his machine measured 10 inches across. (diameter)

What was the volume that the machine could hold?

If the diameter of the sphere is 10 inches, then the radius is 5.
$V_{sphere} = (4/3)\pi r^3 = (4/3)(3)5^3 = 500$ cubic inches.

# Complete Solutions and Answers  203–207

203. Suppose the starting set is the set of all people who are currently students at KITTENS University. Would the rule *Assign to each student their art history teacher* be a function?

    Joe is currently a student at KITTENS University. He is not taking any art history classes. In order for a rule to be a function each member of the starting set must be assigned to exactly one thing. Joe wouldn't be assigned to anything. That rule is not a function.

    When you first learned about fractions we talked about the top and the bottom of a fraction. The grownup words were numerator and denominator.

    With functions, the grownup word for the starting set is called the **domain** of the function.

205. All of these numbers are composite.
    60000000003, 70000011, 8870000001, 45
    What is an easy way to show that none of them are prime?

    They are all divisible by 3.
    (The sum of their digits is divisible by 3.)

206. This is a function. {(a, rat), (b, house), (c, moon)}. What is the domain of this function?

    This function could have been written as      a → rat
                                                       b → house
                                                       c → moon
    The domain (the starting set) is {a, b, c,}.

207. She wanted to create a function which assigned each of her nine pairs of walking shoes to one of the days of the week. The pink pair was assigned to Sunday. The white pair to Monday. The orange pair to Tuesday. The red, to Wednesday. The purple, to Thursday. The blue, to Friday. The brown, to Saturday.
    Was this a function?

    The starting set (the domain) was the nine pairs of shoes. A function is a rule that assigns *to each element* of the domain exactly one answer. Two pairs of her shoes were not assigned. It is not a function.

93

## 209–214 Complete Solutions and Answers

209. It was on special: $148.35 for 23 of them. How much would each fork cost?

$$23 \overline{) 148.35} = 6.45$$

Each fork would cost $6.45.

210. Change to fractions.

$37\% = \dfrac{37}{100}$   $50\% = \dfrac{50}{100} = \dfrac{1}{2}$   $1\% = \dfrac{1}{100}$   $100\% = \dfrac{100}{100} = 1$

212. Using the base 8 numeration system, fill in the blanks.

$613_{\text{base 10}} = \underline{1} \times 8 \times 8 \times 8 + \underline{1} \times 8 \times 8 + \underline{4} \times 8 + \underline{5} \times 1$
$\phantom{613_{\text{base 10}} = \underline{1} \times }512\phantom{\times 8 \times 8 + \underline{1} \times }64\phantom{\times 8 + \underline{4} \times} 8\phantom{+ \underline{5} \times} 1$

$613_{\text{base 10}} = 1145_{\text{base 8}}$

$$\begin{array}{r} 512 \\ 64 \\ 32 \\ +\ 5 \\ \hline 613 \end{array}$$

214. Joe was in his boat, 300 feet from shore. He was heading toward shore at the rate of 5 feet per second.

Once he got to shore, he hopped on his bicycle and peddled at the rate of 25 mph. It was 5 miles from the shore to Darlene's apartment.

How long before Joe was knocking on her door?

$$\dfrac{d}{r} = t$$

The distance, d, is 300 feet. The rate, r, is 5 feet per second. The time it took Joe to reach the shore was $\dfrac{300}{5} = 60$ seconds.

The distance, d, was 5 miles. The rate, r, was 25 mph. The time it took Joe to bicycle to Darlene's was $\dfrac{5}{25} = \dfrac{1}{5}$ hours.

60 seconds + $\dfrac{1}{5}$ hours

= 1 minute + 12 minutes   (the arithmetic: $\dfrac{1}{5}$ of 60 minutes = 12 minutes)

= 13 minutes until Joe was at Darlene's.

# Complete Solutions and Answers | 215–220

**215.** Express 890 as the sum of two primes.

Looking on that list, we notice that 887 is a prime.
890 = 887 + 3   There might be other answers.

**216.** Instead of his usual order of 68 cases, he would order 25% more. How many cases would he order?

We'll skip doing this problem the hard way.

The easier way:   100% + 25% = 125%
125% of 68 cases = 1.25 × 68 = 85
He would order 85 cases each month.

**218.** Joe's $39 jelly bean machine was 30% of the value of all the things Coalback took that night. How much had he taken?

39 is 30% of ?
30% is the number closest to the *of.*   0.3) 39.

$$\phantom{0}130.$$
$$3.)\,390.$$

Coalback has stolen $130 that night.

**220.** KITTENS University income is two million dollars a year. It receives 98% of its income from student tuition. (The rest is received from gifts. It receives no funds from the federal or state governments.)
How many dollars does it receive from tuition?

98% of $2,000,000 = 0.98 × 2,000,000 =

```
  2000000.           zero digits to the right of the decimal
×    0.98            two digits to the right of the decimal
  16000000
  18000000
 196000000.  ⇒ 1960000.   two digits to the right of the decimal
```

KITTENS receives $1,960,000 in tuition each year.

## 222–228 Complete Solutions and Answers

222. Fill in the blanks.

8174. = 8 thousands + 1 hundred + 7 tens + 4 ones
　　　　　10×10×10　　　10×10　　　10　　　1

223. She walks at the rate of 2.4 miles per hour (mph). From the previous problem, you know the radius is 3.6 miles.

$$\frac{d}{r} = t = \frac{3.6}{2.4} = 2.4\overline{)3.6}$$

$$24\overline{)36.0} = 1.5$$

It will take Darlene 1.5 hours to walk out and get the newspaper. And, of course, 1.5 hours to walk back to the house.

225. Multiply each of these by a thousand.

　　2.789　　4　　7.007　　0.0063

　　2789.　　4000.　7007.　　6.3　⎫
　　2789　　4000　　7007　　6.3　⎬ either of these lines is okay

227. To multiply by 10 you move the decimal 1 place to the right.
To multiply by 100 you move the decimal 2 places to the right.
To multiply by 1,000 you move the decimal 3 places to the right.
To multiply by 10,000 you move the decimal 4 places to the right.

　　How many places to the right do you move the decimal if you are multiplying by 1,000,000?

　　If you noticed that the number of places to the right that you move the decimal is the same as the number of zeros after the 1, then this problem is easy.

　　To multiply by 1,000,000 you move the decimal 6 places to the right. 35.9078 multiplied by a million is 35,907,800.

228. If three primes add up to an even number, must one of the primes be equal to 2?

　　All the primes are odd except 2. Three odd numbers always add to an odd answer. So if three primes add to an even number, one of them has to be 2.

## Complete Solutions and Answers | 230–236

**230.** $751_{\text{base 10}}$ = __ × 16×16×16 + __ × 16×16 + __ × 16 + __ × 1
= __ × 4096 + __ × 256 + __ × 16 + __ × 1

Now comes a little arithmetic.
= _0_ × 4096 + _2_ × 256 + __ × 16 + __ × 1

```
  751
−  512
  239
```

= _0_ × 4096 + _2_ × 256 + _14_ × 16 + __ × 1

```
  239
−  224
   15
```

= _0_ × 4096 + _2_ × 256 + _14_ × 16 + _15_ × 1

So $751_{\text{base 10}}$ = 2 14 15$_{\text{base 16}}$ = 2ef$_{\text{base 16}}$

**231.** When the radius of that balloon ball was four feet, how much air was in it?

$V_{\text{sphere}} = (4/3)\pi r^3 \approx (4/3)(3)(4^3)$ = 256 cubic feet

**234.** Express one-sixth of a ton of a potato chip in pounds. Round your answer to the nearest pound.

$\dfrac{1/6 \text{ ton}}{1} \times \dfrac{2000 \text{ pounds}}{1 \text{ ton}}$ = 6)2000.0 (333.3) ≐ 333 pounds

**236.** Joe would get a larger plate—one with an area of 200 square inches. If the cook filled that big plate with jelly beans, each square inch would hold 7.8 jelly beans. How much would the whole plate hold?

As a d = rt problem...
Whole plate = (7.8 beans per square inch) × (200 square inches)
= 1,560 beans

As a conversion factors problem...

$\dfrac{200 \text{ square inches}}{1} \times \dfrac{7.8 \text{ beans}}{1 \text{ square inch}}$ = 1,560 beans

## 237–239 Complete Solutions and Answers

**237.** Convert $\frac{3}{8}$ into a percent.

To convert a fraction into a percent, you first convert it into a decimal. Fractions ⇒ Decimals ⇒ Percents

$\frac{3}{8}$ means

```
      0.375
   8) 3.000
    - 24
      60
    - 56
      40
    - 40
```

0.375 = 37.5% or 37½%

**238.** The fall semester at KITTENS University is 18 weeks long. When Joe had done 3 days of the fall semester, he asked Darlene how long it was before the end of the semester. Darlene couldn't figure it out and asked Betty. Show Betty's computation.

```
   18 weeks                              17 weeks 7 days
 −        3 days     becomes           −         3 days
                                         17 weeks 4 days
```

*Dear Darlene,*

*There are 17 weeks and 4 days left in the semester.*

*—Betty*

**239.** When Joe got to Fred's arithmetic class, he had already eaten three-fourths of a bag. While Fred lectured, he finished that bag and ate 2⅓ pounds in a second bag. How much did Joe eat during Fred's lecture?

If Joe had eaten three-fourths of the first bag, then there was one-fourth of the first bag not yet eaten.

$\frac{1}{4} \times \frac{10}{1} = \frac{10}{4} = 2\frac{2}{4} = 2½$ pounds in the first bag.

2½ pounds in the first bag plus 2⅓ pounds in the second bag:

```
     2½              2 3/6
  +  2⅓              2 2/6
  ─────────         ───────
                     4 5/6   pounds were eaten by Joe during Fred's class
```

98

# Complete Solutions and Answers | 240–247

**240.** Fill in the blanks.

82,507 = __8__ ten thousands + __2__ thousands + __5__ hundreds + __0__ tens + __7__ ones.

**241.** Let R = {eagle, canary, parrot}.
Let S = {eagle, moose}.

A) Is R ⊂ S true?
   No. Not every element of R is an element of S.

B) R – S = {canary, parrot}
   R – S means those elements that are in R but not in S.

C) R ∪ S = {eagle, canary, parrot, moose}
   The order you list these four elements doesn't matter.
   You should not list *eagle* twice.

D) R ∩ S = {eagle}

**242.** 0.611 – 0.4789 = ?

$$\begin{array}{r} 0.611 \\ -\ 0.4789 \end{array} \quad \Rightarrow \quad \begin{array}{r} 0.6110 \\ -\ 0.4789 \\ \hline 0.1321 \end{array}$$

**244.** At $1,000 per acre, how much would those 10,240 acres cost?

$1,000/acre times 10,240 acres = $10,240,000.

(To multiply by a 1,000 move the decimal three places to the right.)

**246.** π = 3.14159265358979323846264338332795....

Round off pi to the nearest hundredth.

3.14**1**59265358979323846 ≐ 3.14

**247.** $46_{\text{base 10}} = ?_{\text{base 2}}$

46 = 1×32 + 0×16 + 1×8 + 1×4 + 1×2 + 0×1 = $101110_{\text{base 2}}$

99

| 250–258 | Complete Solutions and Answers |

250. $45 + $2.98 = ?

```
  45.00
+  2.98
  47.98
```

$45 + $2.98 = $47.98

253. This is a function: {(ball, string), (mouse, 5), (door, Germany)}. What is the set of all first coordinates called?

The set of first coordinates is {ball, mouse, door}. It is called the **domain** of the function (or the starting set).

254. Which of these are true?
  A) California ∈ { x | x is a state in the United States}    True.
  B) 2 ∈ { x | x > 7}   False. 2 isn't a number that is larger than 7.
  C) Fred ∉ {x | x is a teacher at KITTENS University}
       False. Fred is a teacher at KITTENS University.
  D) {15, 16} = {x | x is a natural number and 14 < x < 17}
       True. Two sets are equal if they have the same members. The only two natural numbers between 14 and 17 are 15 and 16. (The natural numbers are 1, 2, 3, 4, 5. . . .)

256. (continuing problem #184) Express $\frac{1}{6}$ as a decimal and express the "leftovers" as a fraction.

```
      0.166666666 4/6
  6) 1.000000000
     6
     ―
     40
     36
     ――
     40
     36
     ――
     etc.
```

$\frac{1}{6} = 0.16666\ 4/6 = 0.16666\ 2/3$

You could have written 0.166 ⅔ or you could have written 0.1666666666666666666 ⅔.

258. $400_{\text{base 10}} = ?_{\text{base 20}}$

$400_{\text{base 10}} = \underline{\phantom{x}} \times 20 \times 20 + \underline{\phantom{x}} \times 20 + \underline{\phantom{x}} \times 1$

$\phantom{400_{\text{base 10}}} = \underline{\ 1\ } \times 20 \times 20 + \underline{\ 0\ } \times 20 + \underline{\ 0\ } \times 1$
$\phantom{400_{\text{base 10}} =\ \ \ \ \ \ \ \ } 400 \phantom{\ \times 20 \times 20\ \ \ \ } 20 \phantom{\ \times 20\ \ \ \ } 1$

$400_{\text{base 10}} = 100_{\text{base 20}}$

# Complete Solutions and Answers | 259–263

**259.** On Tuesday she had used 70% of a bottle. Today, Wednesday, she finished that bottle and used three-eights of a second bottle.

What percent of a quart had she used today?

On Wednesday she used 30% of the first bottle and 37½% of the second bottle. 30% + 37½% = 67½% of a quart. (Knowing that three-eights equals 37½% makes this problem easy.)

**260.** What would be the circumference of a circle whose radius is $\frac{3}{4}$?

If the radius is $\frac{3}{4}$ then the diameter would be $1\frac{1}{2}$

(The arithmetic: $\frac{3}{4} \times 2 = \frac{3}{4} \times \frac{2}{1} = \frac{3}{\cancel{4}_2} \times \frac{\cancel{2}^1}{1} = \frac{3}{2} = 1\frac{1}{2}$)

If the diameter is $1\frac{1}{2}$ then the circumference would be $4\frac{5}{7}$

(The arithmetic: $1\frac{1}{2} \times 3\frac{1}{7} = \frac{3}{2} \times \frac{22}{7} = \frac{3}{\cancel{2}_1} \times \frac{\cancel{22}^{11}}{7} = \frac{33}{7}$

$= 7\overline{)33}\ ^{4\ R\ 5} = 4\frac{5}{7}$ )

**261.** Express 3.2 hours as hours and minutes.

The 3 hours is 3 hours. That's logical.
The 0.2 hours we need to change into minutes.

$\frac{0.2\ hours}{1} \times \frac{60\ minutes}{1\ hour} = 12\ minutes$

↑ conversion factor

3.2 hours = 3 hours, 12 minutes

**263.** $10,000 is what percent of $80,000?

You divide the number closest to the *of* into the other number.

$\frac{10000}{80000} = \frac{1}{8} = 12½\%$     These Nine Conversions can come in handy.

101

| 264–269 | Complete Solutions and Answers

264. Change into dollars.
　　　55¢ = $0.55　　　　744¢ = $7.44　　　　3¢ = $0.03

266. Joe to make 36 easy monthly payments to pay a total of $2,257.92. How much would it cost Joe each month?

```
        62.72
    36)2257.92
       216
        97
        72
       259
       252
         72
         72
          0
```

Joe would make payments of $62.72 each month.

> This would not have bothered Joe at all. He often goes for months without opening his mail. "Too many words" was his favorite description of mail.

268. Joe measured this parallelogram. Is it possible to find the area just using his measurements?

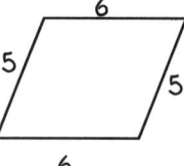

There is not enough information to find the area. The parallelogram might look like this

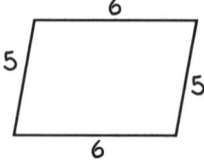

or like this

269. He began at 10:47 A.M. and finished at 1:31 P.M. How long did it take Ralph to create his famous perfume?

　　　From 10:47 A.M. to 12:47 P.M. is 2 hours.
　　　From 12:47 P.M. to 1 P.M. is 13 minutes.
　　　From 1 P.M. to 1:31 P.M. is 31 minutes.
The total elapsed time is 2 hours and 44 minutes.

# Complete Solutions and Answers  270–273

**270.** The main room in Joe's studio apartment measures 12.2' by 14.7'. The ceiling is 8.2'. What is the volume of this room?

```
     14.7        one digit to the right of the decimal
  ×  12.2        one digit to the right of the decimal
     294
     294
     147
    17934   ⟹  179.34    two digits to the right of the decimal
```

```
    179.34
 ×     8.2
    35868
   143472
  1470588   ⟹  1470.588
```

Joe's room has a volume of 1,470.588 cubic feet

**272.** Darlene has 48 bottles of red nail polish, 6 of blue polish, and 2 of yellow polish. Express this as a continued ratio.

48:6:2

**273.** Graph {(2, 1), (3, 12), (4, 2)}.

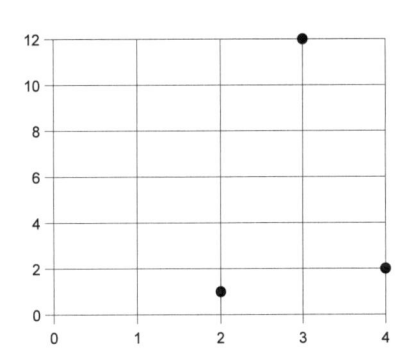

Or you might have drawn:

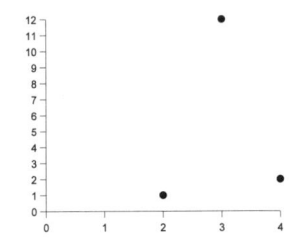

103

| 274–278 | Complete Solutions and Answers |

274. One bag covers 0.04 square miles. We want to cover 2.0 square miles.

Step 1: We know that 1 bag = 0.04 square miles.

Step 2: The conversion factor will be either $\dfrac{0.04 \text{ square miles}}{1 \text{ bag}}$ or it will be $\dfrac{1 \text{ bag}}{0.04 \text{ square miles}}$

Step 3: We are given 2.0 square miles. We want to convert that into bags.
Step 4: We pick the conversion factor (from step 2) so that the units cancel.

$$\frac{2.0 \text{ square miles}}{1} \times \frac{1 \text{ bag}}{0.04 \text{ square miles}}$$

$$= \frac{2.0 \; \cancel{\text{square miles}}}{1} \times \frac{1 \text{ bag}}{0.04 \; \cancel{\text{square miles}}}$$

$$= 0.04 \overline{)2.0} \quad = \quad 4 \overline{)200.}^{\,50.}$$

Darlene will need 50 bags of grass seed.


276. Joe owed $171.36 to the doctor because of his fishing accident. Would it be possible for him to make three equal payments?

Is 171.36 evenly divisible by 3? If we add up the digits (1 + 7 + 1 + 3 + 6) we get 18. Since 18 is evenly divisible by 3, so is 171.36.


278. Instead of walking 3.6 miles out to the brick wall to get the newspaper, Darlene would send her dog, Wolfie. Wolfie could run at 12 mph. How long would it take him to go from the house to the wall?

$$t = \frac{d}{r} = \frac{3.6}{12} = 12\overline{)3.6}^{\,0.3}$$

It would take Wolfie 0.3 hours.

## Complete Solutions and Answers — 280–282

**280.** $751_{\text{base 10}}$ = __ × 8×8×8 + __ × 8×8 + __ × 8 + __ × 1
      = __ × 512 + __ × 64 + __ × 8 + __ × 1

Now comes a little arithmetic.

      = _1_ × 512 + __ × 64 + __ × 8 + __ × 1

```
  751
− 512
  ───
  239
```

      = _1_ × 512 + _3_ × 64 + __ × 8 + __ × 1

```
  239
− 192
  ───
   47
```

      = _1_ × 512 + _3_ × 64 + _5_ × 8 + __ × 1

```
   47
−  40
  ───
    7
```

      = _1_ × 512 + _3_ × 64 + _5_ × 8 + _7_ × 1

So $751_{\text{base 10}}$ = $1357_{\text{base 8}}$


**282.** Joe would get a job and make $1,875/week. Taking Thursdays off would decrease his salary by one-fifth. How much would he be making?

**Hard way:**

He would lose one-fifth of his wages.     ← two steps
$\frac{1}{5} \times 1{,}875 = \$375$.
He would have left $1{,}875 - 375 = \$1{,}500$/week.

**Easier way:**

Taking Thursdays off he would keep four-fifths of his salary.
$\frac{4}{5} \times 1{,}875 = \$1{,}500$/week.

105

| 285-298 | Complete Solutions and Answers

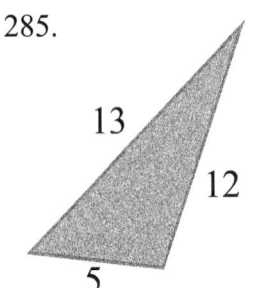

285. When we know the lengths of the sides of a triangle and want to find its area, we use Heron's formula.

perimeter p = 5 + 12 + 13 = 30 meters
semiperimeter s = 30/2 = 15 meters

$$A = \sqrt{s(s-a)(s-b)(s-c)}$$

$$= \sqrt{15(15-5)(15-12)(15-13)}$$

$$= \sqrt{15(10)(3)(2)}$$

$$= \sqrt{900}$$

To find $\sqrt{900}$ we have to do some trial and error guessing.

Try 20.   20 × 20 = 400     Too small
Try 25.   25 × 25 = 625     Still too small.
Try 40.   40 × 40 = 1600    Too large.
Try 30.   30 × 30 = 900     Yes.

$\sqrt{900}$ = 30 square meters of concrete.

292. Round off 3.14159265358979323846264338332795... to the tenths place.

3.14159265358979323846264338332795... ≐ 3.1

294. (continuing problem #199) How far would they be apart if measured in inches. (Use a conversion factor.)

$$\frac{15.7 \text{ feet}}{1} \times \frac{12 \text{ inches}}{1 \text{ foot}} = \frac{15.7 \cancel{\text{feet}}}{1} \times \frac{12 \text{ inches}}{1 \cancel{\text{foot}}} = 188.4 \text{ inches}$$

298. She bought a square piece of land that was just big enough to contain that circle of grass. What would be the length of one side of that square?

Some questions are easy. If the radius is 5, then the diameter is 10. And that is also the length of a side of the square.

106

## Complete Solutions and Answers  300–302

300. After Joe was 2 weeks and 2 days into the fall semester, he asked Betty how long before the end of the semester. Show Betty's computation.

$$\begin{array}{r} 18 \text{ weeks} \\ - \underline{\phantom{0}2 \text{ weeks} \quad 2 \text{ days}} \end{array} \quad \text{becomes} \quad \begin{array}{r} 17 \text{ weeks} \quad 7 \text{ days} \\ - \underline{\phantom{0}2 \text{ weeks} \quad 2 \text{ days}} \\ 15 \text{ weeks} \quad 5 \text{ days} \end{array}$$

Dear Joe,
    There are 15 weeks and 5 days left in the semester.
        —Betty

301. A full blow by Joe was 2 quarts of air.
    The balloon was 256 cubic feet.
    How many blows did Joe do in order to inflate that balloon?
    (A dry quart is equal to about 67 cubic inches. There are $12^3$ cubic inches in a cubic foot. Use conversion factors to convert one balloon into blows by Joe.)

$$\frac{\text{one balloon}}{1} \times \frac{256 \text{ cubic feet}}{1 \text{ balloon}} \times \frac{12^3 \text{ cubic inches}}{1 \text{ cubic foot}} \times \frac{1 \text{ dry quart}}{67 \text{ cubic inches}}$$

$$\times \frac{1 \text{ blow by Joe}}{2 \text{ dry quarts}} = \frac{256 \times 12^3}{67 \times 2} \text{ blows by Joe} \doteq 3{,}301 \text{ blows}$$

Just for fun... If Joe could do a puff every 5 seconds, then

$$\frac{3301 \text{ puffs}}{1} \times \frac{5 \text{ seconds}}{1 \text{ puff}} \times \frac{1 \text{ minute}}{60 \text{ seconds}} \times \frac{1 \text{ hour}}{60 \text{ minutes}} \doteq 4.6 \text{ hours}$$

302. Here is a number with the last digit missing.
    5646868116876168468448465161168746511177777993?
    What could that last digit be in order to easily know that the number is not prime?

    If the last digit were 0, 2, 4, 6, or 8, we would know that the number was even. Even numbers (except 2) are all composite.
    If the last digit were 0 or 5, we would know that the number was divisible by 5.

107

| 304–310 | Complete Solutions and Answers |

304. One of the nicest thing about these black steel forks is that they would stack up.  Five of them made a stack that was 7 cm (cm = centimeter) tall.  How tall was each of them?

```
       1.4
   5) 7.0000
      5
      20
      20
       0
```

Each fork was 1.4 cm tall.
(A centimeter is a little less than half an inch.)

306. Joe stuck all nine fishing flies in the pocket over his heart.  He did that because he knew that fishing was close to his heart.

If the domain is the nine fishing flies, is this a function?

Was each of the flies assigned to exactly one pocket?  Yes.  It is a function.

If each of the flies had a fishing hook in it, this was not a bright place to store his flies.

307. The base of the triangle is 50'.  The height (also known as the altitude) is 20'.

A = ½bh = ½(50)(20) = 500 square feet

308. Express 270 as the sum of two primes.

269 is a prime, but that won't help, since 1 isn't a prime.
263 is a prime, and that will work.
270 = 263 + 7    There might be other answers.

310. It was on sale for $993,250.  She knew that after she and Joe got married they could save $7.25 each week.  How many weeks would it take them to buy the mansion?

This conversion factor will change dollars ($) into weeks.

$$\frac{\$993{,}250}{1} \times \frac{1 \text{ week}}{\$7.25} = 7.25 \overline{)993250.}$$

$$725. \overline{)99325000.} \quad 137000.$$

It will take them 137,000 weeks to save enough to buy the house.

108

# Complete Solutions and Answers  312–317

312. Express 250 as the sum of three primes.

By the previous problem, we know that one of the three primes must be equal to 2.

250 – 2 = 248.

So what two primes add to 248?  Looking at the Handy List of Primes, 241 and 7 will work.

250 = 2 + 241 + 7   There might be other answers.

313. Goldbach's first conjecture was that every even number greater than 2 can be written as the sum of two primes.  Express 46 as the sum of two primes.

46 = 23 + 23

Wait a minute!  46 can also be written as 3 + 43.  Goldbach said that it can be done, but he didn't say that it can be done in only one way.

Wait a minute!  46 = 17 + 29

315. Change to a percent.
0.89 = 89%    0.5 = 0.50 = 50%    0.04 = 4%

316. 42.6 – 3.008 =

```
  42.6           42.600
– 3.008    ⇒   – 3.008
                39.592
```

317. Joe would spent 50% of his time reading the Donald Duck comic, 40% of his time reading the Woody Woodpecker comic, and 10% of his time resting his eyes. Draw a circle graph (a pie chart) showing this.

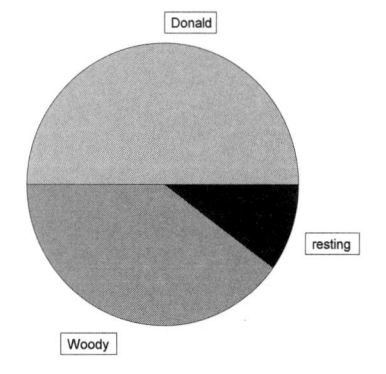

### Joe's Reading Times

(Pie chart with Donald 50%, Woody 40%, resting 10%)

109

| 319–323 | Complete Solutions and Answers |

319. Change $\frac{5}{6}$ into a percent.

$\frac{5}{6}$ means $6\overline{)5.00}$ = 0.83 R 2   I could have written $0.8\overline{33}$ but a remainder works better here.

$\frac{5}{6} = 0.83 \frac{2}{6} = 0.83⅓ = 83⅓\%$

321. He set the price at $1.50, but no one wanted to buy it. Then he put it on sale at two-fifths off of the original price. He then knocked an addition one-third off of the sale price. What was the price after these two discounts were made?

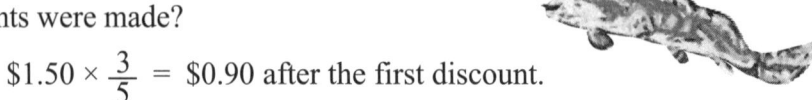

$1.50 × $\frac{3}{5}$ = $0.90 after the first discount.

$0.90 × $\frac{2}{3}$ = $0.60 after the second discount.

$0.60 or 60¢   Either one is correct.

Had Joe offered his fresh fish for 60¢, he might have had a buyer.

323. He decided to sell his $624 stash at a 35% discount to Sergeant Snow. How much would Snow pay him?

35% discount means that Snow would be paying 65%.
65% of $624 ➡ 0.65 × 624

```
      624      zero digits to the right of the decimal
   ×  0.65     two digits to the right of the decimal
     3130
     3744
    40560   ➡   405.60
                   two digits to the right of the decimal
```

Snow would pay $405.60 to Coalback.

# Complete Solutions and Answers | 325–330

**325.** Fill in the blanks.

5.238 = 5 ones + 2 tenths + 3 hundredths + 8 thousandths
          1       1/10      1/100      1/1000

**327.** The current weight was 1,800 pounds.
How many kilograms is that? (one pound ≈ 0.45 kilograms)

Using a conversion factor, $\dfrac{1800 \text{ pounds}}{1} \times \dfrac{0.45 \text{ kilograms}}{1 \text{ pound}}$

= 1800 × 0.45 kilograms = 810 kilograms

**329.** Which of these are functions?
   A) {(hat, rain), (shoe, rain), (glove, rain)}
      Yes.  hat is assigned exactly one answer.
             shoe is assigned exactly one answer.
             glove is assigned exactly one answer.
      That is the definition of a function.
   B) {(7, grape), (7, peach)}
      No.  7 has been assigned two answers.
   C) {(1, 2), (2, 3), (3, 4), (4, 5)}
      Yes.  1 is assigned exactly one answer.
             2 is assigned exactly one answer.
             And so on.
   D) {(Newton, 77), (Archimedes, 77)}
      Yes.  Newton is assigned exactly one answer.
             Archimedes is assigned exactly one answer.

**330.** Express $87_{\text{base 10}}$ in base 20 (the vigesimal number system).

$87_{\text{base 10}} = 47_{\text{base 20}}$

Here is the arithmetic:
   4 × 20
+  7 × 1

     Lincoln's most famous speech began, "Four score and seven years ago. . . ." That sounded a lot better than "Eighty-seven years ago. . . ."

| 332–333 | **Complete Solutions and Answers** |

332. Round off 3.141592653589793... to four decimal digits.

    3.14159̷2653589793... ≐ 3.1416

333. In the **base 20 system** (also known as the vigesimal system)
6419 means $6 \times 20{\times}20{\times}20 + 4 \times 20{\times}20 + 1 \times 20 + 9 \times 1$

712 in the base 20 system equals what in our base 10 system?

$712 = 7 \times 20{\times}20 + 1 \times 20 + 2 \times 1 =$     2800
                                                                                   20
                                                                       +   2
                                                                          2822

So $712_{\text{base 20}} = 2822_{\text{base 10}}$
Numbers are shorter in base 20.

small essay<br>**Why Don't We Use Base 20?**

    The advantage of the vigesimal (base 20) system is that numbers are shorter. If you are a millionaire, you own $\$1{,}000{,}000_{\text{base 10}}$. Expressed in base 20, $\$1{,}000{,}000_{\text{base 10}}$ equals $\$65{,}000_{\text{base 20}}$.

    But there is a big disadvantage in using the vigesimal system.

    In the base 10 system, there are ten digits: 0, 1, 2, 3, 4, 5, 6, 7, 8, and 9.

    In the base 20 system, we need 20 digits.

    Suppose you have $20_{\text{base 10}}$ rabbits. To count them in the base 20 system

$1_{\text{base 20}}\ 2_{\text{base 20}}\ 3_{\text{base 20}}\ 4_{\text{base 20}}\ \ldots \hfill 10_{\text{base 20}}$

the last rabbit would be $10_{\text{base 20}}$ since $10_{\text{base 20}}$ is $20_{\text{base 10}}$.

    Counting in base 20: 1, 2, 3, 4, 5, 6, 7, 8, 9, ?, ?, ?, ?, ?, ?, ?, ?, ?, ?, 10. We need more digits. We have to invent them. They will look funny because they are new to you. When you first saw 1, 2, 3, and so on, they looked funny. Now they don't seem strange. **(continued next page)**

I'm going to invent ten new digits for the base 20 system. The digits now are: 0, 1, 2, 3, 4, 5, 6, 7, 8, 9, ✣, ❡, ❉, ♣, ❀, ✧, ✇, ✠, ◖, and ✮.

Now $18_{\text{base 10}}$ is ◖. And $19_{\text{base 10}}$ is ✮.

Numbers are shorter in base 20, but learning the multiplication tables is a nightmare.

In good old base 10, there were 100 multiplication facts—everything from 0 × 0 up to 9 × 9.

In the base 20 system, there are 400 multiplication facts to learn! You would be in college before you learned them all.

Would you like to do a little division? ❉❉8✠) 3✮0✧❡✣

That's why we don't use the vigesimal system.

end of small essay

334. Round 1,470.588 off to the nearest cubic foot.

1470.588 ≐ 1,471 cubic feet.

335. If Darlene had spent 3 years so far working on getting Joe to propose marriage and if she was 6% of the way toward that goal, how long would the whole process take?

Three years is six percent of the whole process.
3 = 6% of the whole process
We divide the number closest to the *of* into the other number.

0.06) 3.

   50.
6) 300.

It looks like Darlene has a half century (50 years) of work to get Joe to propose.

336. Round each of these to the nearest hundred.

673 ≐ 700     ≐ means "rounded off"

82.0082 ≐ 100

10,501 ≐ 10,500

49 ≐ 0

## 337-339 Complete Solutions and Answers

337. What is the smallest member of this set? {x | x is a whole number that is divisible by 3 and x > 10,040,200}

Is 10,040,201 divisible by 3? No. The sum of the digits = 8.
Is 10,040,202 dividible by 3? Yes. The sum of the digits = 9.

Therefore, the smallest member of {x | x is a whole number that is divisible by 3 and x > 10,040,200} is 10,040,202.

For fun I could list the members of {x | x is a whole number that is divisible by 3 and x > 10,040,200}. They are {10,040,202, 10,040,205, 10,040,208, 10,040,211, . . . }.


339. How much was Fred expecting his $2,300 to turn into?

A thousand times $2,300 is $2,300,000, which is *two million, three hundred thousand dollars*. Not many five-and-a-half-year-olds have this much money.

small essay
### How Did the Duck Fund Do So Well?

The Duck Fund actually did multiply an original investment a thousand fold. Coalback wasn't lying about this. The Duck Fund had one investor, and that person had invested one cent. (1¢ or $0.01)

That one cent had become 1,000¢ (or $10.00).

There are two ways to lie. You can lie by not telling the truth, or you can lie by not telling *the whole truth*.

Here is the whole truth: Coalback had created a zillion funds. There was the Aardvark Fund, Alligator Fund, Ants Fund, Arctic Wolf Fund, Armadillo Fund, Atlas Beetle Fund, Badger Fund, Bees Fund, Beetle Fund, Bengal Tiger Fund, Bison Fund, Blister Beetle Fund, Blue Whale Fund, Blue-footed Booby Bird Fund, Boa Constrictor Snake Fund, Brown Recluse Spider Fund, Butterflies Fund, Camel Fund, Caribou Fund, Cat Fund, Centipede Fund, Cheetah Fund, Chickadee Fund, Chicken Fund, Chimpanzee Fund, Chinchilla Fund, Chipmunk Fund, Cow Fund, Coyote Fund, Crab Fund, Crocodile Fund, Darkling Beetle Fund, Deer Fund, Dingo Fund, Dog Fund, Dolphins Fund, Dragonfly Fund, Duck-billed Platypus Fund, **Duck Fund**, Dung Beetle Fund, Earthworm Fund, Elephant

Fund, Elk Fund, Emu Fund, Giant Panda Bear Fund, Giraffe Fund, Goat Fund, Golden Eagle Fund, Golden Pheasant Fund, Gorilla Fund, Grasshopper Fund, Great White Shark Fund, Grizzly Bear Fund, Guinea Pig Fund, Hamster Fund, Harp Seal Fund, Hedgehog Fund, Hermit Crab Fund, Hippopotamus Fund, Hoopoe Bird Fund, Hornet Fund, Horse Fund, Impala Fund, Jackal Fund, Jaguar Fund, Jellyfish Fund. . . .

What Coalback didn't mention was that all the funds (except the Duck Fund) lost all their money. He had placed wild bets in the stock market. Only one bet had paid off. That was the Duck Fund. Coalback pretended that the other funds never existed.

end of small essay

340. Is *take a natural number and multiply it by 7* a function?

Yes. If I take a number, say 5, and multiply it by 7, I will always get the same answer. Each element in the starting set has a unique answer.

341. Express $\frac{1}{6}$ as a decimal and put a bar over the repeating part.
$\frac{1}{6} = 0.1\overline{6}$

342. `Beef ribs take 500% longer than two minutes to cook on a grill.` What is the total time to cook those ribs?

500% more means 500% + 100%.
600% of 2 minutes = 6. × 2 = 12 minutes to cook the ribs.

Joe thought 500% meant five hundred times longer.

the results of Joe's math error

| 344–348 | Complete Solutions and Answers |

344. Joe's diet and Fred's diet are very different. Joe has 18 cavities in his teeth. Fred has none. Draw a bar graph.

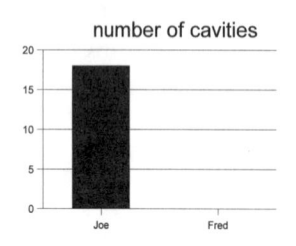

or you could have drawn

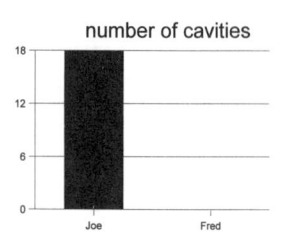

346.   purchase cost         $179.
       tax                   14.32
       wool cutting scissors 38.55
       vet costs             60.
                             ———
                             $291.87

The sheep would cost $291.87. The article failed to mention a lot of other costs. Darlene would have to make sure that there was water for the sheep to drink. She would have to make sure that no lions, tigers, wolves, or neighbors would eat her sheep. (If her mansion were in Kansas, she wouldn't have to worry about lions or tigers or neighbors. I'm not sure about wolves.)

348. The two triangles are congruent.
(Translation: They have the same shape and size.)
The two triangles have the same area.
If I could figure out the area of one of the triangles,
I would double it to find the area of the parallelogram.
   How do I find the area of a triangle if I know
the lengths of the three sides? I use Heron's formula.

$p$ = the perimeter = $10.5 + 13.9 + 7.4$
$s$ = the semiperimeter = $p/2$
Area = $A = \sqrt{s(s-a)(s-b)(s-c)}$

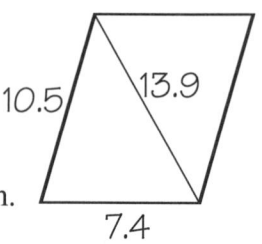

116

# Complete Solutions and Answers  350–352

350. Which of these are subsets of {C, M, V}?

    A) {C}                  Yes. {C} ⊂ {C, M, V}

    B) {M, V}             Yes. {M, V} ⊂ {C, M, V}

    C) {V, M, C}          Yes, because every element of {V, M, C} is also in {C, M, V}

    D) {A, B, C, M, V}    No. Not every element of {A, B, C, M, V} is in {C, M, V}.

352. *Take a number, double it, and then subtract 4* is a function. What is the inverse of that function?

    If we start with, say, 12, and use this function, we will get 20.

(The arithmetic: 12 doubled is 24. Subtract 4 and we get 20.)

    To find the inverse function we need to do things backwards. We first *add 4, and then divide by 2.*

    Applying this inverse function to 20, we will get back our original 12.

(The arithmetic: 20 plus 4 is 24. Divide 24 by 2 and we get 12.)

small essay<br>

**Finding Inverse Functions**

    Once you've done a couple of these, they become sleepy-time easy.

    Suppose the function rule is *Take a number, add 15, multiply by 4, subtract 4000, divide by 67.*

    Then the inverse would be *Multiply by 67, add 4000, divide by 4, and subtract 15.*

    If you *Put your foot into a sock and then put your sock into a shoe,* the inverse would be *Take your sock out of your shoe and then take your foot out of your sock.*

    If you *Put frosting on the cake, then put the cake in a box, and then put the box in your car,* the inverse would be *Take the box out of the car, take the cake out of the box, and then remove the frosting.*

    It is like running a movie in reverse.

end of small essay

| 354–360 | **Complete Solutions and Answers**

354. Just pour an ounce of Ralph's into the water and your drain will open in 45 minutes. Joe's drain had been clogged up with jelly bean gunk. At 3:52 P.M. he added an ounce of Ralph's and waited. When should Joe expect the drain to open?

From 3:52 to 4:00 is 8 minutes.
From 4:00 to 4:37 is 37 minutes.
8 + 37 = 45 minutes. His drain should be open by 4:37.

356. What is the smallest member of this set? {x | x is a whole number that is divisible by 5 and x > 17}

Step 1: The whole numbers are 0, 1, 2, 3, 4, 5, 6. . . .
Step 2: The whole numbers that are divisible by 5 are 0, 5, 10, 15, 20, 25, 30, 35, 40. . . .
Step 3: The whole numbers that are divisible by 5 and are greater than 17 are 20, 25, 30, 35, 40. . . . The smallest member of this set is 20.

357. The CD cost him 90% of the money he had in his pocket ($20). How much did that CD cost?

90% of $20    We know both sides of the *of* so we multiply.
0.90 × 20.

$$\begin{array}{r} 20 \\ \times\ 0.9 \\ \hline 180 \end{array}$$ ➡ 18.0 ➡ $18

358. Graph {(3,1), (2, 2), (1, 4)}.

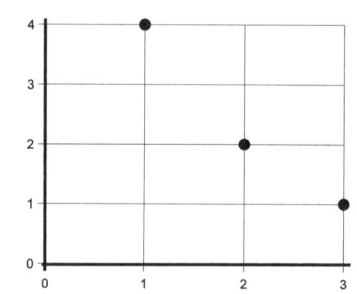

360. If a circle has a diameter of a million kilometers, what is its circumference? (Use 3.1416 for $\pi$.)

C = $\pi$d ≈ 3.1416d = 3.1416 × 1,000,000 = 3,141,600 kilometers

## Complete Solutions and Answers — 362–362

**362.** Next month this vase will be worth 12% more. Each month it will be worth 12% more than the previous month.

The vase was only $10. Darlene sent in her money. It was August. How much would it be worth in September? In October?

<small>First of all, it should be noted that the vase was plastic and only one inch tall, and it didn't look anything like the picture in the ad. In August, Coalback sold 3,000 of these cheap vases and made almost $30,000. He then went out of business and headed to a remote part of Kansas where no one could find him.</small>

If it were worth $10 in August, it would be worth $10 × 1.12 = $11.20 in September. (100% + 12% = 112% = 1.12)

If it were worth $11.20 in September, it would be worth $11.20 × 1.12 = $12.544 ≐ $12.54 in October.

| | |
|---|---|
| Worth in August | 10. |
| Worth in September | 10. × 1.12 |
| Worth in October | 10. × 1.12 × 1.12 |
| Worth in November | 10. × 1.12 × 1.12 × 1.12 = 10 × (1.12)$^3$ |
| Worth in December | 10 × (1.12)$^4$ |

In a year its worth would be  10 × (1.12)$^{12}$

In a decade (10 years = 120 months) its worth would be 10 × (1.12)$^{120}$

What's (1.12)$^{120}$ equal to? The hard way to figure this out would be to multiply 1.12 × 1.12 × 1.12 × 1.12 × 1.12 × 1.12 × 1.12 × 1.12 × 1.12 × 1.12 × 1.12 × 1.12 × 1.12 × 1.12 × 1.12 × 1.12 × 1.12 × 1.12 × 1.12 × 1.12 . . . × 1.12.

In advanced algebra we will learn about logs. Then doing all these multiplications can be done in three easy steps!

| | |
|---|---|
| Take the log of 1.12 | 0.049218 |
| Multiply by 120 | 5.9061627 |
| Take the anti-log | 805,680. |

So, using logs, (1.12)$^{120}$ ≐ 805,680.

In a decade Darlene's $10 vase would be worth $10 × 805,680, which is

# $8,056,800.

Darlene could sell her little plastic vase and buy her mansion and have enough left over to hire a cook and a butler.

| 365–378 | Complete Solutions and Answers |

365. The Egg sign is correct. They didn't mean $255. for a dozen eggs. The Pet Mice sign is wrong. They didn't mean that each mouse would be less than 1¢. (.87 < 1 so .87¢ < 1¢)

The Jelly Bean sign is probably incorrect. .01¢ means that each bean would cost one-hundredth of a penny. Then 1¢ would buy 100 jelly beans. Then $1 would buy 10,000 jelly beans. (To multiply by a hundred you add two more zeros.)

367. What is the largest element of this set? {x | x is a natural number and 44 < x < 66 and x is divisible by 3}

    The natural numbers = {1, 2, 3, 4, 5 . . .}.
    The natural numbers between 44 and 66 are 45, 46, . . . , 64, 65.
    65 is not divisible by 3 since 6 + 5 is not divisible by 3.
    64 is not divisible by 3 since 6 + 4 is not divisible by 3.
    63 is divisible by 3.

    The largest element of the set is 63.

368. There were 15 blue ones, 5 black ones, 5 orange ones, and 75 green ones. Without looking, he popped one of them into his mouth. What is the probability that it was orange?

    There were 100 jelly beans. (15 + 5 + 5 + 75) Five of them were orange. The probability he picked an orange one is $\frac{5}{100}$ which is 5% or $\frac{1}{20}$.

376. One-fortieth of 4816.0 square feeet.

$$40 \overline{)4816.0} = 120.4$$

Joe's shack would be 120.4 square feet.

378. If one box of Banana Jelly Beans in a Box cost $1.04, how much would a million boxes cost?

    To multiply by 1,000,000, you move the decimal 6 places to the right. $1.04 becomes $1,040,000.

    To multiply by a trillion (which is $1,000,000,000,000 = 10^{12}$), you move the decimal 12 places to the right.

## Complete Solutions and Answers — 380–385

380. $0.038 + 0.88 = ?$

$$\begin{array}{r} 0.038 \\ + 0.88 \\ \hline 0.918 \end{array}$$

381. That canoe might be worth $300. (It's not new.) Joe's original dream boat is worth $80,000. What is the ratio of the value of his dream boat to that canoe?

$80,000:300$ or $\dfrac{80,000}{300}$ or $\dfrac{800}{3}$ or $800:3$

Any of these is correct.

---

**English**

It is not "Any of these *are* correct."
"Any of these is correct" is a shortened form of "Any one of these is correct."
The subject of the sentence is *one*, not *these*.

---

383. If there are 5,000 student at KITTENS and the total tuition paid is $1,960,000, what is the tuition that each student pays?

$5000\overline{)1960000.}$ = $392.$     Each student pays $392.

In comparison, in the year this book was written . . .
Stanford University's tuition     $45,729
Harvard     $59,550
Yale     $45,800

384. There were 15 blue ones, 5 black ones, 5 orange ones, and 75 green ones. Without looking, he popped one of them into his mouth. What is the probability that it is either black or green?

There are 80 that are either black or green. The probability is $\dfrac{80}{100}$ or 80% or $\dfrac{4}{5}$.

385. The ratio of sunny days to cloudy days in Kansas is $\dfrac{7}{8}$. Express this ratio using colons.

7:8     or     7 sunny days:8 cloudy days

121

## Complete Solutions and Answers

386. Which of these are true?

A) $3 \in \{x \mid 1 < x \leq 7 \text{ and } x \text{ is a whole number}\}$

True. The whole numbers are $\{0, 1, 2, 3, 4 \ldots\}$.

$\{x \mid 1 < x \leq 7 \text{ and } x \text{ is a whole number}\} = \{2, 3, 4, 5, 6, 7\}$

B) $\{7, 10\} \subset \{x \mid 6 < x < 11 \text{ and } x \text{ is a whole number}\}$

True.

$\{x \mid 6 < x < 11 \text{ and } x \text{ is a whole number}\} = \{7, 8, 9, 10\}$

$\{7, 10\}$ is a subset of $\{7, 8, 9, 10\}$ because every element of $\{7, 10\}$ is an element of $\{7, 8, 9, 10\}$.

C) $5.4 \in \{x \mid x \text{ is a whole number}\}$

False. 5.4 is not a whole number.

D) $\frac{25}{5} \notin \{x \mid \text{ is a natural number}\}$

False.

$\{x \mid \text{ is a natural number}\} = \{1, 2, 3, 4, 5, 6 \ldots\}$

$\frac{25}{5}$ is equal to 5, which is a natural number.

388. Joe filled one of Darlene's vases with 60 pounds of jelly beans.

Three different problems...

A) 7% of those beans were orange. How many pounds of orange jelly beans were in that vase?

7% of 60 pounds = $0.07 \times 60$ = 4.2 pounds were orange.

B) If mice ate 7% of all the jelly beans, what would be the weight of the beans that remained?

(100% − 7% = 93%)

93% of 60 pounds = $0.93 \times 60$ = 55.8 pounds were left.

C) If Joe added 7% more to the original 60 pounds, what would be the new weight?

(100% + 7% = 107%)

107% of 60 = $1.07 \times 60$ = 64.2 pounds would be the new weight.

## Complete Solutions and Answers — 390–398

**390.** 69 × 0.007

```
      69        0 digits to the right of the decimal
   × 0.007      3 digits to the right of the decimal
   ------
     483  ➟  0.483   3 digits to the right of the decimal
```

**393.** The width of the can is 3". Find the area of the top of the can. Round your answer off to the nearest square inch.

If the diameter is 3", then the radius is 1.5".
$A = \pi r^2 \approx 3.1 \times 1.5 \times 1.5 = 6.975$ square inches

$6.975 \doteq 7$ square inches

**395.** A dancing pizza ran 128 miles. How long would it take Joe to run 128 miles? Joe can run at the rate of 8 mph.

$$\frac{d}{r} = t \qquad t = \frac{d}{r} = \frac{128 \text{ miles}}{8 \text{ mph}} = 16 \text{ hours}$$

We divided miles by mph and got an answer of hours. That is not really so strange.

Do you remember how to divide fractions? $\frac{a}{b} \div \frac{c}{d}$

becomes $\frac{a}{b} \times \frac{d}{c}$

In the same way $\frac{\text{miles}}{\text{miles/hr}}$ becomes $\frac{\text{miles}}{1} \div \frac{\text{miles}}{\text{hour}}$

which becomes $\frac{\text{miles}}{1} \times \frac{\text{hour}}{\text{miles}}$

which equals hours.

**398.** Add, using decimals. $\frac{7}{100} + \frac{943}{1000} + \frac{8}{10}$

```
    0.07
    0.943
  + 0.8
  -------
    1.813
```

123

### 399–403 Complete Solutions and Answers

399. Convert 137,000 weeks into years (assuming 50 weeks in a year).

$$\frac{137000 \text{ weeks}}{1} \times \frac{1 \text{ year}}{50 \text{ weeks}} = 50\overline{)137000} = 2740$$

It will take Darlene and Joe 2,740 years to save enough to buy that mansion.

Darlene had to do some figuring. First, Joe would have to give up fishing. Then she would give up wearing lipstick. Then they could save $19.86 each week. Every 52 weeks they would save $1,032.72. Then it would take only about 960 years to buy the house.

Giving up eating, they could buy the house in 400 years, which is four centuries.

It never occurred to Darlene that making more money might be the easier approach. Of course, making more money would mean that they would have to offer the world something more valuable than being able to eat jelly beans or put on nail polish.

400. 0.7 means the same thing as 0.70. To show that this is true, change both 0.7 and 0.70 into fractions and show that they are equal.

$$0.7 = \frac{7}{10} \qquad 0.70 = \frac{70}{100}$$

We notice that $\frac{70}{100}$ reduces to $\frac{7}{10}$ when you divide top and bottom by 10.

402. (continuing problem #360) Round that circumference off to the nearest million kilometers.

3,141,600 kilometers ≐ 3,000,000 kilometers.

≐ means rounded off to

≈ means roughly equal to

403. There were 15 blue ones, 5 black ones, 5 orange ones, and 75 green ones. What is the probability that the jelly bean won't be either black or green?

There are 20 jelly beans (out of a 100) that aren't black or green.

The probability that the jelly bean won't be either black or green is $\frac{20}{100}$ which is 20% or $\frac{1}{5}$

## Complete Solutions and Answers  404–410

**404.** Change to a decimal.

*The hard way:*

$71\% = \frac{71}{100} = 0.71 \qquad 6\% = \frac{6}{100} = 0.06 \qquad 3.02\% = \frac{3.02}{100} = 0.0302$

*The easier way:*

$71\% = 0.71 \qquad\qquad 6\% = 0.06 \qquad\qquad 3.02\% = 0.0302$

Just move the decimal two places to the left.

**405.** The gym would be 35 yards by 80 yards. What is the area of this rectangle?

$A_{rectangle} = lw = 35 \times 80 = 2{,}800$ square yards

**406.** Four pounds of Freedonia jelly beans occupied one cubic foot. The dump truck delivered 6,000 pounds to Joe's front door. Using a conversion factor, how many cubic feet were in that delivery?

We want to convert 6,000 pounds into cubic feet. We start with the 6,000 pounds.

$$\frac{6000 \text{ pounds}}{1} \times \frac{1 \text{ cubic foot}}{4 \text{ pounds}} = \frac{6000 \cancel{\text{ pounds}}}{1} \times \frac{1 \text{ cubic foot}}{4 \cancel{\text{ pounds}}}$$

= 1,500 cubic feet.

Since the main room in Joe's apartment is 1,471 cubic feet, he will have to store some of the Freedonia jelly beans in his bathroom.

**408.** Joe had packed 38,394 lemon jelly beans for his fishing trip. Could he divide them equally into five pockets of his new fishing jacket?

No. 38,394 is not evenly divisible by 5 (because the last digit, 4, is not a 0 or a 5).

**410.** She would buy 400 vases. Thirty percent of them would be porcelain. How many would that be?

30% of 400 = 0.3 × 400 = 120 vases would be porcelain.

125

| 411–419 | Complete Solutions and Answers

411. Fred would eventually get back one-hundredth of his $2,300 investment.

How much did Fred get back?

One-hundredth of $2,300 is $23.00.

413. In Fred's geometry class he once noticed that 24 students were wearing white socks and 36 students were wearing blue socks. He said that the ratio of white-sock wearers to blue-sock wearers was 2:3.

Explain how he got that answer.

The ratio is 24:36. This is the same as $\frac{24}{36}$

This fraction reduces to $\frac{2}{3}$ and this is the ratio 2:3.

416. A box of banana jelly beans weighs 5 pounds 3 ounces. (1 lb. = 16 oz.) Joe opened a box and ate 7 ounces. How much did that box now weigh?

|  5 lbs. 3 oz. | 4 lbs. 16 oz. + 3 oz. | 4 lbs. 19 oz. |
| − 7 oz. | − 7 oz. | − 7 oz. |
| | | 4 lbs. 12 oz. |

After Joe ate 7 ounces, the box weighed 4 pounds 12 ounces.

417. Darlene has a lot of nail polish bottles. If she picks one at random, there is a 7% chance that it will be blue. What is the chance that it won't be blue?

If there is a 7% that it will be blue, there is a 93% (100% − 7%) chance that it won't be blue.

419. Joe filled 11 lunch bags with jelly beans. Each bag contained 807 jelly beans. Name three numbers that divide evenly into the number of jelly beans Joe consumed.

1 will divide evenly into any whole number.

He ate 11 × 807 jelly beans. Both 11 and 807 divide evenly into 11 × 807.

## Complete Solutions and Answers | 421–429

**421.** Those 23 forks cost $170 when you include the shipping, the handling, and the taxes. They could be paid for in "8 easy monthly payments." How much would Darlene pay each month for all these wonderful forks?

```
         21.25
      8)170.0000
        16
        ‾‾
         10
          8
          ‾
          20
          16
          ‾‾
           40
           40
           ‾‾
            0
```

For eight monthly payments of $21.25 Darlene could own these black steel forks.

**423.** {(1,3), (5, 2), (7, 7)}

**425.** 20 white, 5 black, 10 spotted

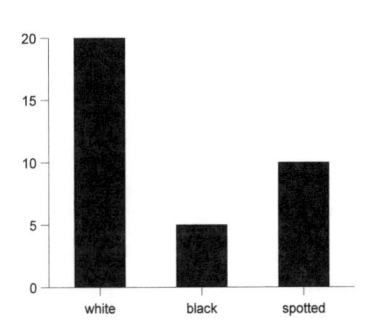

**427.** What is the radius of the biggest circle that could fit inside of the square plot of land?

*Don't you wish that all the problems were this easy?*

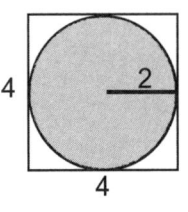

**429.** Twenty-one jelly beans occupy 6 cubic inches. We want to convert 500 cubic inches into jelly beans.

$$\frac{500 \text{ cubic inches}}{1} \times \frac{21 \text{ jelly beans}}{6 \text{ cubic inches}} = 1{,}750 \text{ jelly beans}$$

## 430–444 Complete Solutions and Answers

**430.** If there is a 100% probability that Joe won't propose to Darlene this week, what is the probability that he will propose?

100% probability = certainty. He will not propose.
The probability that he will propose is 0%.

**436.** 0.05 × 0.09

```
    0.05      2 digits to the right of the decimal
  × 0.09      2 digits to the right of the decimal
    ────
     45  ⟹  0.0045    4 digits to the right of the decimal
```

**440.** (continuing problem #327) How much will it cost Joe to get Jelly Bean shipped to him?

We knew that Jelly Bean weighs 810 kilograms.
We know that $3.87 = 1 kilogram.

Using a conversion factor, $\dfrac{810 \text{ kilograms}}{1} \times \dfrac{\$3.87}{1 \text{ kilogram}}$

= 810 × $3.87 =

```
     3.87      2 digits to the right of the decimal
   ×  810      0 digits to the right of the decimal
   ─────
     000
     387
    3096
   ──────
   313470  ⟹  3134.70    2 digits to the right of the decimal
```

The shipping cost will be $3,134.70

**442.** What is the area of the parallelogram?

A = base times height = bh
  = 40 × 50 = 2000 square feet

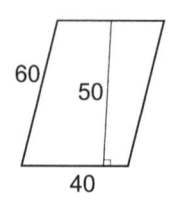

**444.** There were 8,200,000 grains of rice. Would he be able to give everyone the same number of grains?

Is 8,200,000 evenly divisible by 3? The sum of the digits of 8,200,000 is 10. Since 10 is not divisible by 3, neither is 8,200,000.

# Complete Solutions and Answers    446–450

446. The can is 6 inches tall. (h = 6".) Its radius is 1.5". Find the volume of the can.

$$V = \pi r^2 h \approx 3.1 \times 1.5 \times 1.5 \times 6 = 41.85 \text{ cubic inches.}$$

If length is measured in inches, then
>area is measured in square inches and
>>volume is measured in cubic inches.

If length is measured in meters, then
>area is measured in square meters and
>>volume is measured in cubic meters.

If length is measured in yards, then
>area is measured in square yards and
>>volume is measured in square yards.

Etc.

448. If the steel flag weighs 5 pounds per square foot, how much would the 500 square foot flag weigh?

This problem could be done in two different ways.

First way. Using d = rt—
the weight would equal 5 pounds per square foot times 500 square feet, which is 2,500 pounds.

Second way. Using a conversion factor, converting square feet into pounds—

$$\frac{500 \text{ square feet}}{1} \times \frac{5 \text{ pounds}}{1 \text{ square foot}} = 2{,}500 \text{ pounds}$$

450. Joe is 5' 6". Jelly Bean is 75' 4". How much taller is Jelly Bean than Joe?

```
   75'  4"                 74' 12"+4"                74' 16"
 −  5'  6"    becomes    −  5'    6"    becomes    −  5'  6"
                                                     69' 10"
```

Jelly Bean is 69' 10" taller than Joe.

129

| 452–458 | Complete Solutions and Answers |

452. Name the subsets of {2, 9}.

They are { }, {2}, {9}, and {2, 9}.

454. π = 3.14159265358979323846264338327 95. . . .
Round π off to the nearest thousandths place.

π ≐ 3.142        ≐ means "rounded off."

455. The probability that the next fish Joe catches is a tuna is $\frac{3}{80}$. The probability that Fred will own a car sometime in the next ten years is $\frac{20}{21}$. What is the probability of both of these things happening?

Those two events are not related. The probability is $\frac{3}{80} \times \frac{20}{21}$

$= \frac{\cancel{3}^1}{\cancel{80}_4} \times \frac{\cancel{20}^1}{\cancel{21}_7} = \frac{1}{28}$

456. Suppose the domain is the set of teachers at KITTENS University. Would the rule *Assign to each teacher the day of the week that they were born on* be a function?

Yes. Everybody was born on some day of the week. And nobody was born on two different days of the week. Each teacher is assigned exactly one day of the week.

458. Complete each of these.
A) {moose, rat, fish} ∩ {rat, 6, A} = {rat}
The intersection, ∩, are those things in both sets.

B) {moose, rat, fish} – {rat, 6, A} = {moose, fish}
The difference, –, are those things that are in the first set, but not in the second set.

C) {x | x is an animal} ∪ {x | x is a dog} = {x | x is an animal}
Since every dog is an animal, the union, ∪, is simply the set of all animals.

D) {x | x is a star in the sky} ∩ Ø = Ø
Ø is the empty set. Ø = { }.

# Complete Solutions and Answers | 461–463

461. Now it's your turn. Change 16⅔% into a fraction.

$$16\tfrac{2}{3}\% = \frac{16\tfrac{2}{3}}{100}$$

Dividing by a 100 is the same as moving the decimal two places to the left.

$$= 16\tfrac{2}{3} \div 100$$

$$= \frac{50}{3} \div 100$$

When you divide (or multiply) mixed numbers, you have to first change them into improper fractions. Changing 16⅔ into an improper fraction means "3 times 16 . . . plus 2"

$$= \frac{50}{3} \div \frac{100}{1}$$

$$= \frac{50}{3} \times \frac{1}{100}$$

$$= \frac{1}{6}$$

462. How do you multiply 3% times 7%?

It's easy . . . if you first turn them into decimals. $0.03 \times 0.07 = 0.0021$, which is 0.21%.

463. Which of these are true?
   A) ∅ = { }     True. ∅ is just another name for { }.
   B) { } ⊂ {2, 18, 444}     True. { } would *not* be a subset of {2, 18, 444} if you could find a member of { } that wasn't in {2, 18, 444}.

   Stated a different way: Every member of { } is a member of {2, 18, 444}. The fancy name for this is that it is *vacuously true*.

   In the same way, it is true to say that every 500-pound cow in your pocket is named Harvey. If you could find a 500-pound cow in your pocket, I would bet a million dollars that its name is Harvey.

   This is fun. Every girl who is older than her mother is a jar of peanut butter. You find me such a girl, and I guarantee that she's a jar of peanut butter. Since you can't find such a girl, what I'm saying is true.

   Every jar of peanut butter that is made out of toenails tastes good.

131

| 466–473 | Complete Solutions and Answers |

466. {(boat, dish), (dish, soap), (sky, rope)} is a function. What is the image of dish under this function?

    This function is the same as:
        boat → dish
        dish → soap
        sky → rope

    dish is mapped to soap under this function. The image of dish is soap.

470. Again using $C = (3\frac{1}{7}) \times d$ find the circumference of a circle whose diameter is $4\frac{5}{11}$

    If the diameter is $4\frac{5}{11}$ then the circumference is 14

(The arithmetic: $4\frac{5}{11} \times 3\frac{1}{7}$ = *when you multiply mixed numbers, you have to first change them into improper fractions* $= \frac{49}{11} \times \frac{22}{7} = \frac{\cancel{49}^7}{\cancel{11}_1} \times \frac{\cancel{22}^2}{\cancel{7}_1} = 14$

471. 50% more than 88 means 150% (50% + 100%) times 88.

    $1.5 \times 88 = 132$

    The only problem with a piano having 132 keys is that there is no music written for 132 different notes.

473. If Joe writes 33 words, he will usually erase 7 of them. If he writes 264 words, how many will he erase?

    We want to convert 264 written words into erased words.

$$\frac{264 \text{ written words}}{1} \times \frac{7 \text{ erased words}}{33 \text{ written words}}$$

$$= \frac{264 \text{ \cancel{written words}}}{1} \times \frac{7 \text{ erased words}}{33 \text{ \cancel{written words}}}$$

$$= \frac{264 \times 7}{33} \text{ erased words} = 56 \text{ erased words}$$

132

# Complete Solutions and Answers | 475–490

**475.** While fishing, Joe eats jelly beans at the rate of 27 per minute. How many will he eat in 3 hours? (First, change hours into minutes.)

$$\frac{3 \text{ hours}}{1} \times \frac{60 \text{ minutes}}{1 \text{ hour}} = 180 \text{ minutes}.$$

*There are two ways to look at this problem.*

**First way:** You can think of it as a d = rt problem.

$$d = rt = 27 \text{ jelly beans per minute} \times 180 \text{ minutes} = 4{,}860 \text{ jelly beans}$$

**Second way:** You can think of it as a conversion factor problem in which you change minutes into jelly beans.

$$\frac{180 \text{ minutes}}{1} \times \frac{27 \text{ jelly beans}}{1 \text{ minute}} = 4{,}860 \text{ jelly beans}$$

Both ways give the same answer.

**478.** Express $\frac{5}{6}$ as a decimal. Put a bar over the repeating part.

$$6 \overline{) 5.000} = 0.833 \qquad \frac{5}{6} = 0.8\overline{3} \qquad \text{or you could have written } 0.83\overline{3}$$

**480.** His rent was 1 hundred + 2 tens + 3 ones + 4 dimes (tenths) + 3 cents (hundredths), which is $123.43.

**484.** ✔ Syrup Soup. It comes in a 9" tall can with a radius of 2".
✔ Sugar Soup. It comes in a 4" tall can with a radius of 3".

$$V_{\text{Syrup Soup}} = \pi r^2 h = \pi \times 2^2 \times 9 = 36\pi$$
$$V_{\text{Sugar Soup}} = \pi r^2 h = \pi \times 3^2 \times 4 = 36\pi$$

The two cans have the same volume.

**490.** Round each of these off to the nearest hundred.

$$772 \doteq 800 \qquad 8225 \doteq 8200 \qquad 61.0086 \doteq 100 \qquad 05.9 \doteq 0$$

## 492–500 Complete Solutions and Answers

**492.** One is what percent of seven? Round your answer to the nearest percent.

1 = ?% of 7
7 is the closest number to *of*.

$$\begin{array}{r} 0.142 \\ 7\overline{)1.000} \\ -\underline{7} \phantom{00} \\ 30 \phantom{0} \\ -\underline{28} \phantom{0} \\ 20 \\ -\underline{14} \end{array}$$

1 is 14.2% of 7
14.2% ≐ 14%

**494.** (continuing problem #294) If Joe were to take some 8-inch (diameter) plates and line them up between himself and Darlene, how many plates could he fit across the 188.4-inch table?

$$\begin{array}{r} 23.5 \\ 8\overline{)188.4} \\ \underline{16} \phantom{00} \\ 28 \phantom{0} \\ \underline{24} \phantom{0} \\ 44 \\ \underline{40} \\ 4 \end{array}$$

So Joe could get at most 23 plates between them.

There is no need to keep dividing. Joe couldn't put part of a plate on the table.

**496.** Fred's salary is $500/month. He donates 10% of his salary to the Sunday school offering. How much is that?

10% of $500 = 0.1 × 500 = 50. (To multiply by 0.1 you move the decimal one place to the left. It is the same as dividing by 10.)

Fred donates $50/month. He is the only kid in Sunday school who donates by check.

**500.** Name the 8 subsets of {a, b, c}.

They are { }, {a}, {b}, {c}, {a, b}, {a, c}, {b, c}, and {a, b, c}.

# Complete Solutions and Answers  |  502–511

**502.** Which one of these is composite?

57, 71, or 101

Fifty-seven is divisible by 3 since 5 + 7 = 12 is divisible by 3.

**504.** What is the circumference of a circular lake that is 0.01 miles across?

C = πd ≈ 3.1 × 0.01 = 0.031 miles.

**506.** Joe weighed 170 pounds. He lost 10% of his weight. He looked in a mirror and thought that he was starting to look a little too skinny. He rushed to the nearest vending machines and emptied them and filled himself. He increased his weight by 10%. How much did he now weigh after losing 10% and then gaining 10%?

Losing 10% of his 170 pounds is like a sale of "10% off."
170 × 0.9 = 153    His skinny weight was 153 pounds.

Gaining 10% from 153 pounds is 100% + 10% = 110%
153 × 1.1 = 168.3 pounds.

After losing 10% and gaining 10% he now weighed 168.3 pounds.

**508.** Joe had $100 in his checking account. He mailed Max a check for $97.99. How much did he have left?

```
   100           100.00
 − 97.99        − 97.99
                   2.01
```

Joe had $2.01 left in his checking account.

**511.** He could hit 11 keys in 15 seconds. How long would it take him to hit all 88 keys?

We want to convert keys into seconds. A conversion factor problem.

$$\frac{\overset{8}{\cancel{88}} \text{ keys}}{1} \times \frac{15 \text{ seconds}}{\underset{1}{\cancel{11 \text{ keys}}}} = 120 \text{ seconds} \quad (= 2 \text{ minutes})$$

135

| 513–517 | **Complete Solutions and Answers** |

513. The union of two sets is abbreviated as ∪. Can you think of a way to remember that?

    The way that I remember that is ∪nion. (There might be other ways. This is just the one I use.)

515. {(boat, dish), (dish, soap), (sky, rope)} is a function.

    Is the set of second coordinates {dish, soap, rope} the domain of that function?

    No. That would be silly. The set of *first coordinates* {boat, dish, sky} is the domain.

    In advanced math courses, we will call the set of second coordinates of a function the **range** of the function. You are too young to learn that now, so we won't mention that fact.

516. At the rally there was a 218.4 pound jelly bean. It was the largest jelly bean that Joe had ever seen. That giant jelly bean was shared equally by the 600 people at the rally. How much did Joe get?

```
        0.364
600) 218.40000
     1800
      3840
      3600
       2400
       2400
          0
```

Each received 0.364 pounds.

517. On June 1, Joe had 4,567 pounds of jelly beans stored at his apartment. By the end of the month he had eaten 100% of them. How many pounds had he eaten?

    100% of 4,567 pounds = 1. × 4567 = 4567

    Joe had eaten 4,567 pounds.

One hundred percent of anything means all of it.

# Complete Solutions and Answers | 519–525

519. Joe had packed 38,394 lemon jelly beans for his fishing trip. Could he divide them equally into two pockets?

    Yes. 38,394 is an even number. Translation: 38,394 is evenly divisible by 2. Its last digit is 0, 2, 4, 6, or 8.

521. A banana jelly bean box measures 8.3" by 7.9" by 10". What is its volume?

```
      8.3
   ×  7.9
     ───
      747
     581
     ────
     6557  →  65.57 square inches.
```

65.57 × 10 = 655.7 cubic inches

523. If the starting set (the domain) is {Fred, Darlene, Joe, Betty, Alexander} and the rule is *Assign to each of these persons the number of jelly beans they purchased in the first 12 hours that Joe's machine was installed on the third floor.* Is this a function?

    Yes. In those first 12 hours, Fred bought none. Darlene → 1. Joe → 439. Betty → 0. Alexander → 0.

    Each element of the domain was assigned exactly one answer—and that is the definition of a function.

    That night after everyone else was asleep, Coalback was on the third floor of the Math Building. He picked up Joe's jelly bean machine and carried in off into the night.

    Have you ever noticed how BIG vending machines are? They are huge and heavy so that people like Coalback won't mess with them.

525. Money poured in for Coalback after that newspaper article. Fred had sent in $2300. Other people had sent in $8.68, $74.99, and $930. How much had these four people sent in?

```
      2300.
         8.68
        74.99
   +   930.
      ──────
      3313.67      Coalback had received $3,313.67.
```

137

| 528–532 | **Complete Solutions and Answers**

**528.** Suppose the domain is {Fred, Alexander, Betty, Joe, Darlene}. Is the rule *Assign to each member of the domain the color of the jelly bean that they have most recently eaten* a function?

If Fred had never eaten a jelly bean, then this wouldn't be a function since a function must assign to each member of the domain exactly one answer.

If all five people in the domain had eaten at least one jelly bean in their life, then it would be a function.

The other reason why that rule might not be a function is that Joe often throws handfuls of jelly beans into his mouth. His current favorite flavors are green fish-flavored jelly beans and pink cherry-marshmallow jelly beans. If he most recently ate both of these at the same time, this rule *the color of the jelly bean most recently eaten* would have two answers: green and pink. For a rule to be a function, each member of the starting set (the domain) must have exactly one answer—not two answers.

**530.** The police are chasing Coalback down an alley. Coalback is running at the rate of 6.7 feet per second.

There is a door 53.6 feet from him. If he can get to that door, he can open it and hide. How long will it take him to reach that door? (Recall $d = rt$ and $\frac{d}{r} = t$)

The distance, d, is 53.6 feet.

The rate, r, is 6.7 feet per second.

$$t = \frac{d}{r} = \frac{53.6}{6.7} = 6.7 \overline{) 53.6} \qquad 67 \overline{) \begin{array}{r} 8 \\ 536 \\ -536 \end{array}}$$

It will take Coalback 8 seconds to reach the door.

**531.** 41.85 cubic inches rounded to the nearest cubic inch is 42 cubic inches.

**532.** Thirty-two of the 88 keys now had purple marks on them. What percent of the keys were marked by Joe? Round your answer to the nearest percent.

$$32 \text{ is ?\% of } 88 \qquad 88 \overline{) \begin{array}{r} 0.36363636 \\ 32.00000000 \end{array}} \qquad 36.\overline{36}\% \doteq 36\%$$

138

# Complete Solutions and Answers | 533–540

**533.** The floor (a rectangle) would measure 20.1' × 45.3'. How many square feet is that?

```
        20.1              1 digit to the right of the decimal
      × 45.3              1 digit to the right of the decimal
       ────
        603
       1005
        804
       ─────
      91053    ⟹  910.53  2 digits to the right of the decimal
```

If I had to order wooden tiles to cover that floor and if each of the tiles was one foot square, I would order 911 of them.

**536.** Suppose the starting set (the domain) is the set of all 5-bedroom houses within 30 miles of KITTENS University. Would this rule be a function? *Assign to each of those houses either Yes or No depending on whether Darlene and Joe could afford to buy it on their current income.*

Yes. That rule is a function. In fact, it is a very easy function. Each of the 5-bedroom houses would be assigned to *No*. A rule is a function if each member of the domain is assigned to exactly one answer.

**538.** Express $\frac{1}{12}$ as a decimal and round your answer to the nearest hundredth.

```
         0.0833
    12)1.0000
       - 96
         ──
          40
        - 36
          ──
           40
         - 36
           ──
          etc.
```

$\frac{1}{12} \approx 0.0833 \doteq 0.08$

**540.** Suppose A = {4, 5, 6} and B = {5, 6, 88}.
Find A ∪ B, A ∩ B, and A − B.

A ∪ B = {4, 5, 6} ∪ {5, 6, 88} = {4, 5, 6, 88}

You don't write {4, 5, 5, 6, 6, 88} because you don't repeat the same element twice in a set.

A ∩ B = {4, 5, 6} ∩ {5, 6, 88} = {5, 6}
A − B = {4, 5, 6} − {5, 6, 88} = {4}

| 542–547 | **Complete Solutions and Answers**

542. Hardwood floors for a gym are expensive. They can cost $70 per square yard. How much will the flooring cost for Darlene's gym?

There are two ways to attack this problem.

First way—as a $d = rt$ problem:

Cost = price per square yard times number of square yards
   = $70/sq yd × 2800 sq yds = $196,000

Second way—using a conversion factor:

$$\frac{2800 \text{ sq yds}}{1} \times \frac{\$70}{1 \text{ sq yd}} = \$196,000$$

544. Joe sat down to listen to Fred's lecture on fractions. It was an hour lecture. After 7 seconds Joe realized that he hadn't brought any jelly beans to eat during the lecture. He would have to go without jelly beans until the end of the lecture. How long would Joe have to suffer?

```
        1 hour
  −               7 seconds

       59 minutes 60 seconds
  −               7 seconds
       59 minutes 53 seconds
```

Joe would have to wait 59 minutes 53 seconds.

547. Suppose the starting set (the domain) is the set of all pairs of whole numbers. Is this rule a function? *Assign to each pair of whole numbers their sum.*

To see if that rule is a function, we have to answer two questions:
① Will you always get an answer when you add two numbers?
② Will you always get just one answer (and not two answers) when you add two numbers?

Since the answer to both questions is Yes, this is a function. This function has a name. It's called addition.

Multiplication is a function. I'm assuming that the starting set is the set of whole numbers. If the domain were the set of all flowers, then it would not be a function since 🌹 × 🌼 doesn't have an answer.

## Complete Solutions and Answers | 549–567

**549.** (continuing problem #516) Using a conversion factor, determine how many ounces of that giant jelly bean Joe received.

$$\frac{0.364 \text{ pounds}}{1} \times \frac{16 \text{ ounces}}{1 \text{ pound}} = 0.364 \times 16 \text{ ounces} =$$

```
    0.364
  ×    16
    2184
    364
    5824   → 5.824 ounces
```

**553.** Change $\frac{3}{8}$ to a percent.

$$8 \overline{) 3.000} \quad 0.375 = 37.5\%$$
(quotient 0.375)

**557.** Joe once wondered how long it would take him to eat a million jelly beans if he were eating them at the rate of 27 per minute. Round your answer off to the nearest minute.

*There are two ways to look at this problem.*

**First way:** As a $\frac{d}{r} = t$ problem.

$$\frac{d}{r} = t = \frac{1,000,000 \text{ jelly beans}}{27 \text{ jelly beans/minute}} \doteq 37,037 \text{ minutes}$$

**Second way:** As a conversion factor problem—converting a million jelly beans into minutes.

$$\frac{1,000,000 \text{ jelly beans}}{1} \times \frac{1 \text{ minute}}{27 \text{ jelly beans}} \doteq 37,037 \text{ minutes}$$

(conversion factor)

**567.** Add, using decimals. $\frac{7}{10} + \frac{6}{100} + \frac{3}{100} + \frac{8}{100} + \frac{5}{1000}$

```
        0.7
        0.06
        0.03
        0.08
  +     0.005
        0.875
```

141

| 569–581 | Complete Solutions and Answers

569.  $V_{sphere} = (4/3)\pi r^3 = (4/3)(3.1)(40)^3 = \dfrac{4}{3} \times \dfrac{3.1}{1} \times \dfrac{64000}{1}$

$= \dfrac{793600}{3} = 264{,}533⅓$ cubic feet.

570. At 9:58 A.M. Joe fell asleep under one of the pianos. At 2:30 P.M. Joe woke up. How long had he slept?

    From 9:58 A.M. to 10 A.M. is 2 minutes.
    From 10 A.M. to 2 P.M. is 4 hours.
    From 2 P.M. to 2:30 P.M. is 30 minutes.

He had slept 4 hours and 32 minutes.

577. Eight beans weighed 5.7 grams. How much would 56 avocado jelly beans weigh?

$\dfrac{56 \text{ beans}}{1} \times \dfrac{5.7 \text{ grams}}{8 \text{ beans}} = \dfrac{\overset{7}{\cancel{56 \text{ beans}}}}{1} \times \dfrac{5.7 \text{ grams}}{\underset{1}{\cancel{8 \text{ beans}}}} = 39.9$ grams

The arithmetic:
```
    5.7
  ×   7
    399   ⟹   39.9
```

The 56 beans would weigh 39.9 grams.

579. A track between himself and Darlene is 188.4 inches long. If the train went 3.1 inches per second, would it reach Darlene in a minute?

$D = rt = \dfrac{3.1 \text{ inches}}{\text{second}} \times \dfrac{60 \text{ seconds}}{1} = 3.1 \text{ inches} \times 60$

```
    3.1      1 digit to the right of the decimal
  × 60       0 digits to the right of the decimal
    00
   186
   1860  ⟹   186.0 inches.   1 digit to the right of the decimal
```

Since 186 < 188.4, the salt would not reach Darlene in a minute.

581. Is 583 closer to 500 or closer to 600?

```
|----------------------------------------------------------|
500                                           583        600
```

It's closer to 600.

142

# Complete Solutions and Answers | 583–590

583. $10,000, but it was on sale for "Five percent off."

**The hard way:**

First, find out how much you save.

5% of $10,000 = 0.05 × 10,000 = $500.

Second, subtract the savings to get the sale price.

10,000 − 500 = $9,500

**The easier way:**

If you are saving 5%, you will be spending 95%.

95% of 10,000 = 0.95 × 10,000 = $9,500

585. If the diameter of Darlene's corral is 1.1 miles, what will the length of the fence for that corral be? (Use 3.1 for $\pi$.)

$C = \pi d \approx (3.1)(1.1)$

```
    3.1
  × 1.1
   ---
     31
    31
   ----
    341   ⟹  3.41
```

The length of the fence will be approximately 3.41 miles.

586. A new pencil might weigh 20 grams. After Joe got through sharpening, it might weigh 1.07 grams. How much was lost?

```
    20          20.00
  − 1.07      − 1.07
   -----       ------
                18.93
```

18.93 grams of pencil were lost.

588. Looking at the bar graphs, we see that Joe uses 2 ounces per month, and Darlene uses 3 ounces per month. Darlene uses more.

590. What is the area of a circle with radius equal to 2 miles.

$A = \pi r^2 \approx 3.1 \times 2 \times 2 = 12.4$ square miles.

Just for fun, notice that the circle with area equal to 12.4 square miles sits inside a square with area 16 square miles.

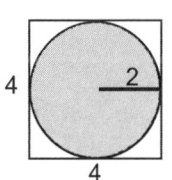

If we had found that the circle had a larger area than the square, we probably had made an arithmetic error.

| 591–595 | **Complete Solutions and Answers** |

591. Express $\frac{1}{6}$ as a decimal and round your answer to the nearest thousandth.

$$0.16666666\ldots \doteq 0.167$$

593. So far in Coalback's life, he had spent 20 years in prison. He's now 36 years old. What percent of his life has he spent behind bars? Round your answer to the nearest percent.

Twenty is what percent of 36?

20 is ?% of 36

We divide the number closest to the *of* into the other number.

```
        0.555
   36) 20.000
      - 180
        200
      - 180
        200
      - 180
```

20 is 55.5% of 36

55.5% $\doteq$ 56%

At some point, because Coalback is a repeat offender, they may stick him in prison for the rest of his life. Then the percent of his life that he will have spent behind bars will climb. He will think back to the few years of his life when he was free.

595. Graph {(1, 200), (2, 600), (3, 100)}. You probably won't use the same scale for the horizontal x-axis and the vertical y-axis.

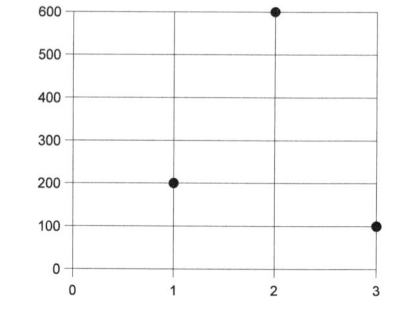

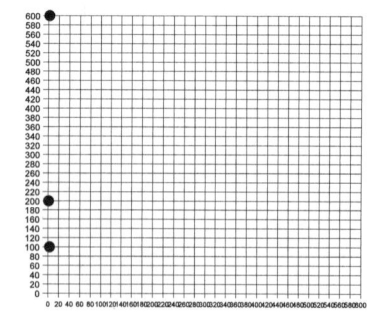

If you had used the same scale for both the x-axis and the y-axis, this is what the graph would look like.

144

## Complete Solutions and Answers | 600–607

**600.** The shipping cost is $3,134.70. On the phone Joe learned that the cost of the He-Man Pet is $8,887. How much is the total that Joe will have to pay?

```
    3134.70
 +  8887.
   12021.70
```
Jelly Bean will cost Joe $12,021.70

**602.** At the rally Joe was awarded one of the 333-pound potato chips. It measured 4.9 feet in its largest dimension.

He decided to build a fence around his chip so that people wouldn't accidentally walk on it and break it. To the nearest tenth of a foot how long would his fence be? (Use 3.1 for $\pi$.)

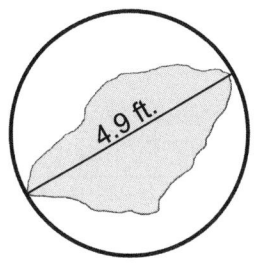

The diameter, d, is 4.9 feet. Circumference, C, is equal to $\pi$d.
C = $\pi$d ≈ 3.1d = 3.1 × 4.9 = 15.19 ≐ 15.2 feet.
( ≈ means "approximately equal to."  ≐ means "rounded off.")

**606.** Syrup Soup is 98% sugar. Express that as a fraction.

$\frac{98}{100} = \frac{49}{50}$

**607.** Coalback never reaches the door. A policeman catches him. We know that $r_{coalback} = 6.7$ feet per second.

Which one of these must be true?

A) $r_{cop} < r_{coalback}$
B) $r_{cop} = r_{coalback}$
C) $r_{cop} > r_{coalback}$

If the policeman catches Coalback, he must have been running faster than Coalback.  C) $r_{cop} > r_{coalback}$ is true.

## 609–613 Complete Solutions and Answers

609. Complete each of these.

A) {dog, cat} ∪ {cat, mouse} = {dog, cat, mouse}  It is considered a misspelling if you list the same member twice. You don't write {dog, cat, cat, mouse.}

B) {Shelly} ∪ {Shelly} = {Shelly}

C) { } ∪ {K} = {K}

611. With the same domain as the previous problem, Would this rule be a function? *Assign to each of those 5-bedroom houses the number of fireplaces it has.*

Yes. Some of the houses would be assigned to zero. Some to one. Some to two. Some to three. Each house would have exactly one answer. That's the definition of a function. (Translation: Every house would be assigned to a number, and no house would be assigned to two different numbers.)

613. It would take him 222 minutes.

A) Convert that to hours and minutes.

$$\frac{222 \text{ minutes}}{1} \times \frac{1 \text{ hour}}{60 \text{ minutes}} = \frac{222}{60} = 60 \overline{)222} \quad 3 \text{ R}42$$
$$\underline{180}$$
$$42$$

222 minutes = 3 hours 42 minutes

B) Convert 222 minutes to hours.

There are two possible answers.
First, as a mixed number.

$$60 \overline{)222} \quad 3 \text{ R}42$$
$$\underline{180}$$
$$42$$

$$222 \text{ minutes} = 3 \frac{42}{60} = 3 \frac{7}{10}$$

Second, as a decimal.

$$60 \overline{)222.0} \quad 3.7$$
$$\underline{180}$$
$$420$$
$$\underline{420}$$

222 minutes = 3.7 hours

## Complete Solutions and Answers | 615–630

**615.** Darlene dreamed about dancing pizzas for 6 minutes. Fred dreamed about dancing pizzas for 12 minutes. Joe, for 3 minutes.

    Draw a bar graph.

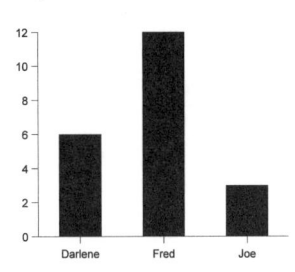

**620.** After they had lain there for 6 days, she noticed that the first fork had rusted. How long would it take for all 23 forks to rust?

    If you multiplied 6 × 23, you did a silly.

    The second fork didn't say to the first fork, "Hey. You rust first, and then I'll start rusting."

    All 23 forks wore a lovely rust-brown coating after six days.

**624.** The starting set is {m, n, p}.

    m → blue

    red is the image of n

    p is mapped to white

    Write this function as a set of ordered pairs.

    {(m, blue), (n, red), (p, white)}

**630.** How many feet would 0.031 miles be?

$$\frac{0.031 \text{ miles}}{1} \times \frac{5280 \text{ feet}}{1 \text{ mile}} = 163.68 \text{ feet}$$

    A sheep would walk 163.68 feet to go around Joe's lake.

640–643 | Complete Solutions and Answers

640. Show that 0.85 and 0.850 are equal by changing them both into fractions and then showing that the fractions are equal.

$$0.85 = \frac{85}{100} \quad \text{and} \quad 0.850 = \frac{850}{1000}$$

The fraction $\frac{850}{1000}$ reduces to $\frac{85}{100}$

if you divide top and bottom by 10.

641. What was the area of her scarf?

(triangle with sides 8.4, base 14.4, and height 6.2)

$$\begin{aligned}\text{Area}_{\text{triangle}} &= (½) \text{ base times height} \\ &= ½\, bh \\ &= ½ \times 14.4 \times 6.2 \\ &= 44.64\end{aligned}$$

643. He took an empty 6-liter Sluice bottle and found that he could fill it with the water that leaked in 4 minutes. He knew that if the boat leaked 198 liters, it would sink.

How many minutes would it take Joe's boat to sink?

Step 1: You know that 6-liters = 4 minutes.
Step 2: The conversion factor will be either $\frac{6 \text{ liters}}{4 \text{ minutes}}$ or $\frac{4 \text{ minutes}}{6 \text{ liters}}$
Step 3: You start with 198 liters. You want to convert that into minutes.
Step 4: You pick the conversion factor (from step 2) so that the units will cancel.

$$\frac{198 \text{ liters}}{1} \times \frac{4 \text{ minutes}}{6 \text{ liters}} = \frac{198 \,\cancel{\text{liters}}}{1} \times \frac{\overset{2}{\cancel{4}} \text{ minutes}}{\underset{3}{\cancel{6}} \,\cancel{\text{liters}}}$$

$$= \frac{198 \times 2 \text{ minutes}}{3} = 132 \text{ minutes}$$

The boat will gain 198 liters of water and sink in 132 minutes.

148

## Complete Solutions and Answers | 645–649

**645.** One month Darlene spent one-eighth of her income on nail polish, one-eighth on lipstick, one quarter on her hair, and one half on clothes.

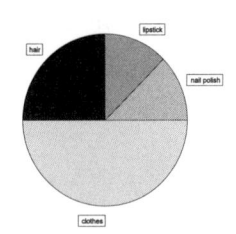

**647.** Expressing $\frac{3}{11}$ as a decimal can be done in several ways.

  A) Express the "leftovers" as a fraction.   Choice #3

$$11 \overline{)3.0000} \quad 0.2727 \; 3/11$$
$$\underline{22}$$
$$80$$
$$\underline{77}$$
$$30$$
$$\underline{22}$$
$$80$$
$$\underline{77}$$
$$3$$

$\frac{3}{11} = 0.2727 \frac{3}{11}$

Or you could have written $0.27 \frac{3}{11}$
or $0.2727272727 \frac{3}{11}$

  B) Use a bar over the repeating part.   Choice #4

$\frac{3}{11} = 0.\overline{27}$   or $0.27\overline{27}$   or $0.272727\overline{27}$

  C) Round your answer to the nearest tenth.   Choice #5

$\frac{3}{11} = 0.2\overline{7} \doteq 0.3$

**649.** Convert 2,500 pounds into tons. (1 ton = 2,000 pounds)

$$\frac{2500 \text{ pounds}}{1} \times \frac{1 \text{ ton}}{2000 \text{ pounds}} = \frac{2500 \text{ tons}}{2000} = \frac{5}{4} \text{ tons} = 1\frac{1}{4} \text{ tons}$$

| 651–662 | Complete Solutions and Answers

651. Joe ate 5 ounces out of the 3-ton shipment. How much was left?

```
      3 tons
−                              5 ounces
_____


      2 tons     2000 pounds
−                              5 ounces
_____


      2 tons     1999 pounds   16 ounces
−                              5 ounces
_____
      2 tons     1999 pounds   11 ounces
```

After Joe ate 5 ounces, there was 2 tons, 1,999 pounds, 11 ounces left.

656. Find the sum of 7.8, 84.09, and 0.885.

$$\begin{array}{r} 7.8 \\ 84.09 \\ +\ 0.885 \\ \hline 92.775 \end{array}$$

658. 42 cubic inches is equal to how many quarts? A cubic inch is approximately equal to 0.017 quarts.

A conversion factors problem. We are converting cubic inches into quarts.

$$\frac{42 \text{ cubic inches}}{1} \times \frac{0.017 \text{ quarts}}{1 \text{ cubic inch}} = 42 \times 0.017 \text{ quarts} = 0.714 \text{ quarts}$$

662. Suppose the domain is {❀,✉,✈} and the rule is ❀ → Fred and ✉ → Betty.

(Often "→" means "is assigned to.")
Why is this rule not a function?

A function must assign to each member of the domain exactly one answer. ✈ didn't get assigned to anything.

## Complete Solutions and Answers | 663–668

663. Round your answer (655.7 cubic inches) off to the nearest cubic inch.
655.7 cubic inches ≐ 656 cubic inches


665. Darlene had spent 4 hours doing three things: reading the ad, painting her nails, and watching television.

Painting her nails took 1 hour and 40 minutes.
Watching television took 2 hours.
How long did she spend reading the ad?

Painting her nails and watching televison took 3 hours and 40 minutes.  Arithmetic:
```
   1 hour and 40 minutes
 + 2 hours
   ─────────────────────
   3 hours and 40 minutes
```

```
    4 hours                    3 hours   60 minutes
 −  3 hours   40 minutes    −  3 hours   40 minutes
                                         20 minutes
```

She spent 20 minutes reading the ad.


668. That graph corresponds to the set of ordered pairs {(1, 3), (2, 2), (2, 7), (4, 4), (5, 1)}. In arrow notation that would be:

$1 \to 3$
$2 \to 2$
$2 \to 7$
$4 \to 4$
$5 \to 1$

2 is in the starting set (the domain). It is mapped to two different elements in the range—both 2 and 7.

The definition of a function is *A rule that maps each element to exactly one member of the range—each element has exactly one answer.*

This is not a function.

151

## 670–677 Complete Solutions and Answers

670. The doctor bill was $605. His student health insurance paid $433.64. Joe had to pay the rest. How much was that?

    605.              605.00
− 433.64       − 433.64
                       171.36

Joe owed $171.36.

672. Create a function with domain = {Joe, Darlene} and with answers in the set {egg, rice}.

One possible function . . .    Joe → egg
                                 Darlene → egg

Second possible function . . .    Joe → egg
                                    Darlene → rice

Third possible function . . .    Joe → rice
                                 Darlene → egg

Fourth possible function . . .    Joe → rice
                                    Darlene → rice

673. Complete each of these.

A) {red, orange, blue} ∩ {orange, blue} = {orange, blue}
B) {red, orange, blue} − {orange, blue} = {red}
C) {x | x is either a cat or a dog} − {x | x is a cat}
      = {x | x is a dog}
D) {x | x is an odd number} ∩ {x | x is an even number} = ∅
    Or you could have written { }.

677. (continuing the problem #557) How many hours would it take Joe? Again, round your answer off to the nearest hour.

$$\frac{37{,}037 \text{ minutes}}{1} \times \frac{1 \text{ hour}}{60 \text{ minutes}} \doteq 617 \text{ hours}$$

# Complete Solutions and Answers | 678–685

678. The total cost of Jelly Bean is $12,021.70. Joe has 90¢ saved up. (90¢ = $0.90) How much more will he need to save in order to get his pet?

$$\begin{array}{r} 12021.70 \\ -\phantom{0000}0.90 \\ \hline 12020.80 \end{array}$$    Joe will need $12,020.80 more.

680. Suppose the starting set (the domain) is the set of all pairs of natural numbers. For example, (3, 88) or (554, 7777777777) or (209, 2).

    Is this rule a function? *Assign to each pair of natural numbers the answer when you divide the first number by the second.*

    Yes. It is a function. Whenever I divide one natural number by another I always get exactly one answer.

    (The domain and the set of answers do not have to be the same set. Two chapters ago we described the function that assigned to each 5-bedroom house within 30 miles of KITTENS University the answer of either Yes or No. Yes is not a house.)

685. Fred is 5½ years old. Let's call it 66 months. He has been teaching at KITTENS University for about 86% of his life. How many months has he been teaching? Round your answer to the nearest month.

    86% of 66 months = 0.86 × 66 = 56.76 ≐ 57
        ( ≐ means rounded off )

    Fred has been teaching for 57 months (out of his 66 months of life.)

    In *Life of Fred: Calculus Expanded Edition* you will read about Fred's birth and how he began teaching at KITTENS at the age of 9 months.

| 688–697 | Complete Solutions and Answers

688. Off in a corner of the gym Darlene would have the workers paint a circle. It would be 40 inches across. That would be a perfect spot for Darlene to play marbles with her girl friends. What would be the area of that circle? (Use 3.1 for $\pi$.)

If the diameter, d, of that circle is 40", then the radius, r, is 20".

$A_{circle} = \pi r^2 \approx 3.1 \times 20 \times 20 = 1{,}240$ square inches

690. $500 + 0.500 = ?$

```
    500.
+   0.500
---------
  500.500
```

694. Is it okay to skip putting in the numbers (the scale) on the x-axis and the y-axis?

No. Those numbers must be there. Without them you might have a graph that looks like . . .

You couldn't tell whether the ordered pairs were (1, 3), (5, 2), and (7, 6) or were (50, 6), (270, 5), and (432, 9).

697.
1. 2 grams of cherry flavor
2. 3.8 grams of marshmallow   } 5.8 grams
3. the rest is sugar.

```
  70,000              70,000.0
-     5.8          −       5.8
                   -----------
                     69,994.2
```

To make 70,000 grams of cherry-marshmallow jelly beans, you need 69,994.2 grams of sugar.

# Complete Solutions and Answers | 700–711

**700.** (continuing #585 in which we learned that we needed 3.41 miles of fencing) If fencing for the corral costs $1000 per mile, how much will the fencing cost for the whole corral?

$1000 \times 3.41 = 3410$. The whole fence will cost $3,410.

Joe didn't much care for horses. He asked Darlene if he could have a pet to keep at their mansion.

She said, "Yes, but only one pet. I want plenty of room for my horses."

Joe said, "Sure. I'll call him Jelly Bean."

J.B.

**701.** $5.827 \doteq 6$ ounces

**705.** (continuing problem #643) First, how many hours and minutes will that be? Second, how many hours will that be?

First, 132 minutes = $60\overline{)132}$ 2 R 12 = 2 hours and 12 minutes

Second, 132 minutes = $60\overline{)132}$ 2 R 12 = $2\frac{12}{60} = 2\frac{1}{5}$ hours

Joe would wait for two hours so that he could fish as long as possible. Then he would row as fast as he could to get to shore. He often made it.

**709.** What is another name for this set? {x | x is a banana jelly bean that came in one of the boxes that Joe bought and x weighs 14,000 tons.}

There aren't any banana jelly beans in any of the boxes that Joe bought that weigh 14,000 tons. This set is called the empty set. It can be written as { } or as ∅.

**711.** Round 163.68 feet to the nearest foot.

$163.68 \doteq 164$ feet

| 714–727 | **Complete Solutions and Answers**

**714.** (continuing the problem #677) If Joe "worked" at eating jelly beans for eight hours each day, how many days would it take him to eat a million jelly beans?

$$\frac{617 \text{ hours}}{1} \times \frac{\text{eight-hour work day}}{8 \text{ hours}} = 77 \text{ work days}$$

Let me do this conversion problem s l o w l y .

Step 1: We know that an eight-hour work day = 8 hours.

Step 2: The conversion factor will be either $\frac{\text{eight-hour work day}}{8 \text{ hours}}$ or it will be $\frac{8 \text{ hours}}{\text{eight-hour work day}}$

Step 3: We are given 617 hours.

Step 4: We pick the conversion factor so that the hours cancel.

$$\frac{617 \cancel{\text{ hours}}}{1} \times \frac{\text{eight-hour work day}}{8 \cancel{\text{ hours}}} = 77 \text{ eight-hour work days}$$

**716.** Fred's trig class has 600 students. Thirty of them use a fountain pen to take notes. What fraction of his class uses a fountain pen?

$\frac{30}{600}$ which reduces to $\frac{3}{60}$ which reduces to $\frac{1}{20}$

**720.** $9.49 + $0 = $9.49      Adding zero to any number doesn't change it.

In algebra, we will say x + 0 = x is true for all x.

**727.** Is this rule a function? *Assign to each pair of whole numbers the answer when you divide the first number by the second.*

Will you always get exactly one answer? No. You can't divide by zero. So (9, 0) won't have an answer. With the domain equal to the whole numbers, this is not a function.

# Complete Solutions and Answers | 731–750

**731.** Syrup Soup is 98% sugar and 2% food coloring.

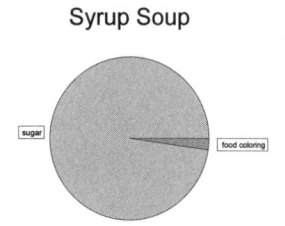

Syrup Soup

[pie chart showing sugar and food coloring]

**734.** Change $4.35 into cents.
Change 4.35 into a percent.

$4.35 = 435.¢     or 435¢
4.35 = 435.%     or 435%

**738.** A) Is 70000000000050000006 divisible by 3?
      Yes, because 7 + 5 + 6 = 18 is divisible by 3.
B) Is it divisible by 5?
      No, because the last digit, 6, is not 0 or 5.
C) Is it divisible by 2?
      Yes, because 6 is divisible by 2.

**740.** Express $\frac{7}{8}$ as a decimal.

$$8\overline{)7.000} = 0.875 \qquad \frac{7}{8} = 0.875$$

**746.** Joe currently saves $0 each week. How much would he save in a century?

    Zero times a zillion weeks is zero. Zero times any number is always zero.
    In algebra we say that for any x,    $0x = 0$

**750.** 67.2 means the same thing as 67.200. To show that this is true, change both into fractions and show that they are equal.

$$67.2 = 67\frac{2}{10} \qquad 67.200 = 67\frac{200}{1000}$$

$\frac{200}{1000}$ reduces to $\frac{2}{10}$ when you divide numerator and denominator by 100.

## 754–766 Complete Solutions and Answers

**754.** Convert $8\frac{3}{8}$ into a decimal.

$$8\overline{)3.000} \quad \frac{0.375}{} \quad \frac{3}{8} = 0.375 \quad \text{and so} \quad 8\frac{3}{8} = 8.375$$

**759.** Suppose the domain is {✿, ✉, ✈} and the rule is ✿ → A and ✉ → C and ✈ → L and ✉ → R. Is this rule a function?

No. For a rule to be a function, each member of the domain must be assigned to ONE thing. ✉ was assigned to both C and R.

**761.** The times that they dreamed were 6, 12, and 3. Which of these numbers are composite?

6 is composite. $6 = 2 \times 3$
12 is composite. $12 = 3 \times 4$
3 is prime. The only divisors of 3 are 1 and 3.

**764.** Let's say that the sixteen digits in the base 16 system are 0, 1, 2, 3, 4, 5, 6, 7, 8, 9, a, b, c, d, e, f.

What does b × c equal in the base 16 system?

In base 16 → 0, 1, 2, 3, 4, 5, 6, 7, 8, 9, a, b, c, d, e, f.
In base 10 → 0, 1, 2, 3, 4, 5, 6, 7, 8, 9, 10, 11, 12, 13, 14, 15.

So b × c in base 16 is the same as 11 × 12 in base 10. $11 \times 12 = 132$.

Now to convert $132_{\text{base 10}}$ into base 16.
132 = __?__ 16's + __?__ 1's.
132 = __8__ 16's + __4__ 1's     *after doing a bunch of arithmetic.*
So b × c = $132_{\text{base 10}}$ = $84_{\text{base 16}}$

**766.** Fill in the blanks.

53.814 = 5 tens + 3 ones + 8 tenths + 1 hundredth + 4 thousandths

# Complete Solutions and Answers | 777–800

777. Subtract 0.294 from 0.85.

$$\begin{array}{r} 0.85 \\ -0.294 \end{array} \text{ becomes } \begin{array}{r} 0.850 \\ -0.294 \\ \hline 0.556 \end{array}$$

785. Ten times 346.5 square feet is 3465 square feet. To multiply by ten you move the decimal one place to the right.

794. Four hours is what fraction of 24 hours?

Easy question. $\frac{4}{24}$ which reduces to $\frac{1}{6}$

798.

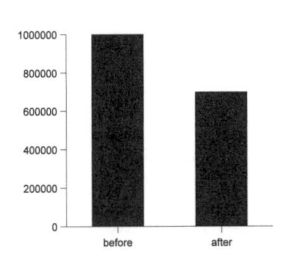

800. Six gallons of Sluice. Seven ounces had leaked out on the way home to her apartment.

$$\begin{array}{r} 6 \text{ gallons} \\ - \quad 7 \text{ ounces} \end{array} \quad - \begin{array}{r} 5 \text{ gallons } 4 \text{ quarts} \\ 7 \text{ ounces} \end{array} \quad - \begin{array}{r} 5 \text{ gallons } 3 \text{ quarts } 32 \text{ ounces} \\ 7 \text{ ounces} \\ \hline 5 \text{ gallons } 3 \text{ quarts } 25 \text{ ounces} \end{array}$$

Darlene had 5 gallons, 3 quarts, 25 ounces of Sluice left for her dinner party.

159

## 806–810 Complete Solutions and Answers

**806.** How long will it take a sheep to run 164 feet at the rate of 8 feet per second?

$$\frac{d}{r} = t$$

A) $\dfrac{164 \text{ feet}}{8 \text{ feet/sec}}$ = $8\overline{)164.0}$ → $20.5$

The sheep will take 20.5 seconds.

or... B) $\dfrac{164 \text{ feet}}{8 \text{ feet/sec}}$ = $8\overline{)164}$ → 20 R4

$\quad\quad\quad\quad\quad\quad\quad\quad$ 16
$\quad\quad\quad\quad\quad\quad\quad\quad$ 04
$\quad\quad\quad\quad\quad\quad\quad\quad\quad$ 0
$\quad\quad\quad\quad\quad\quad\quad\quad\quad$ 4

$20\,\dfrac{4}{8} = 20\,\dfrac{1}{2}$ seconds

**810.** Joe's pet, whom he'd name Jelly Bean, would eat 4.6 tons of food every 6 days. How much would J.B. eat in 33 days?

Step 1: You know that 4.6 tons = 6 days.

Step 2: The conversion factor will be either $\dfrac{4.6 \text{ tons}}{6 \text{ days}}$ or $\dfrac{6 \text{ days}}{4.6 \text{ tons}}$

The conversion factor is equal to one because the top and bottom of the fraction are equal to each other.

Step 3: We are given 33 days, which we want to convert into tons of food.

Step 4: We pick the conversion factor (from step 2) so that the units cancel.

$$\frac{33 \text{ days}}{1} \times \frac{4.6 \text{ tons}}{6 \text{ days}} = \frac{33 \times 4.6 \text{ tons}}{6} = \frac{151.8 \text{ tons}}{6}$$

$6\overline{)151.8}$ → $25.3$
$\quad\underline{12}$
$\quad\;\;31$
$\quad\;\;\underline{30}$
$\quad\quad\;18$
$\quad\quad\;\underline{18}$
$\quad\quad\quad 0$

J.B. will eat 25.3 tons in 33 days.

# Complete Solutions and Answers | 813–819

813. If you add two prime numbers, will the answer always be composite?

*Almost always* the sum of two prime numbers will be composite. That is true because almost all prime numbers are odd. If you add two odd numbers, the answer will be even, and almost every even number is composite.

The big exception is the number 2. It is the only even prime number. Sometimes when you add 2 to another prime you get a prime answer.

For example, 2 + 3 = 5 and 5 is prime.
2 + 5 = 7 and 7 is prime.
2 + 7 = 9 but 9 is not prime.
2 + 11 = 13 and 13 is prime
2 + 13 = 15 but 15 is not prime.

815. When the salt shaker left Joe's side of the table, it contained 60.2 grams of salt. On the train it fell over and leaked salt. During the trip it lost 59.48 grams of salt. How much arrived at Darlene's?

$$\begin{array}{r} 60.2 \\ -\ 59.48 \\ \hline \end{array} \qquad \begin{array}{r} 60.20 \\ -\ 59.48 \\ \hline 0.72 \end{array} \qquad 0.72 \text{ grams of salt arrived.}$$

819. With the same domain as the previous problem, Would this rule be a function? *Assign to each of those 5-bedroom houses the color Red if the house has an even number of windows and the color Blue if the house has two bathrooms.*

No. Some houses would not have any answer. For example, a house with 17 windows and one bathroom. Some houses would be assigned to both Red and Blue. For example, a house with 14 windows and two bathrooms.

To be a function, each element of the starting set (the domain) must have exactly one answer.

161

**821–840** | **Complete Solutions and Answers**

821. The circumference was 31 meters long. What was the radius of that circle?

To go from circumference to diameter, you divide by π.

$$3.1 \overline{)31.} \quad \text{becomes} \quad 31 \overline{)310.}^{\phantom{0}10.}$$

The diameter is 10 meters. The radius is 5 meters.

830. He paid ten dollars for it, which included a sales tax of $0.75. How much was the price of the backpack before the sales tax was added?

$10. − $0.75

```
   10              10.00
−   0. 75    ⇨   −  0.75
                   ─────
                    9.25
```

The price of the backpack was $9.25.

836. Is this a function? *Assign the fish to ☺ if I catch you today, and assign the fish to ☹ if I don't catch you today.*

Every fish will either be caught or it won't be caught. It is a function.

838. The case measured 12" by 12" by 10". What is its volume? (In algebra we will write $V_{box} = \ell w h$.)

$V_{box} = \ell w h = 12 \times 12 \times 10 = 1440$ cubic inches.

840. What fraction of a cherry-marshmallow jelly bean is sugar?

In the previous problem we found that there are 69,994.2 grams of sugar in 70,000 grams of jelly bean.

Therefore, $\frac{69994.2}{70000}$ of the jelly bean is sugar.

## Complete Solutions and Answers | 844–852

844. Express $\frac{312}{999}$ as a decimal.

$$\frac{312}{999} = 999 \overline{)312.000000000000}^{\phantom{0}0.312312312312 \text{ R } 312}$$

$$\phantom{999)}\underline{2997}$$
$$\phantom{9999)}1230$$
$$\phantom{9999)}\underline{999}$$
$$\phantom{9999)}2310$$
$$\phantom{9999)}\underline{1998}$$
$$\phantom{9999)}3120$$
$$\phantom{9999)}\text{etc.}$$

Using "leftovers" as a fraction (Choice #3)  $\frac{312}{999} = 0.312312 \frac{312}{999}$

or $0.312 \frac{312}{999}$  or  $0.312312312312 \frac{312}{999}$

Using a bar over the repeating part (Choice #4) $\frac{312}{999} = 0.\overline{312}$

846. If Coalback can run 6.7 feet per second and the policeman can run $8\frac{3}{8}$ feet per second, Coalback's speed is what percent of the policeman's speed?

 6.7 is what percent of $8\frac{3}{8}$

 6.7 is ?% of 8.375      $8\frac{3}{8} = 8.375$ by the previous problem

 8.375 is closest to the *of*.    $8.375\overline{)6.7}$

$$8375.\overline{)6700.0}^{\phantom{0}0.8}$$
$$\phantom{8375.)}\underline{-67000}$$

Coalback can run 80% of the policeman's speed.


852. In our base 10 decimal system $731 = 7 \times 100 + 3 \times 10 + 1 \times 1$. What does 00731 equal?

$00731 = 0 \times 10000 + 0 \times 1000 + 7 \times 100 + 3 \times 10 + 1 \times 1$, which, of course, equals $7 \times 100 + 3 \times 10 + 1 \times 1$.

That's why $00731 = 731$. ☺

## 854–860 Complete Solutions and Answers

854. Is 800000000000000000005 evenly divisible by 3?

   No. The sum of the digits is 13, and 13 is not divisible by 3.

858. Fill in one word: {4, 7} is a  subset  of {4, 5, 7, 8, 9}.

860. Banana jelly beans are just like regular bananas. They come in three colors: green, yellow, and brown. Each box contains all three colors.

   Joe opened one of his boxes of banana jelly beans. Let those beans be the domain (the starting set). Is this rule a function? *Joe put the green beans in the refrigerator. He put the brown beans in the garbage.*

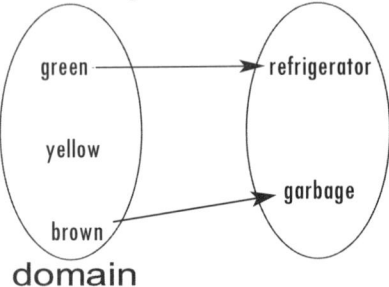

```
   green ─────────► refrigerator

   yellow
                    garbage
   brown ──────────►
   domain
```

   The yellow beans are not assigned to anything. A function is a rule that assigns *to each element of the domain* exactly one answer. This is not a function.

164

# Complete Solutions and Answers | 870–872

870. Express $\frac{1}{6}$

  A) as a decimal

I have several choices.

✓ Express the "leftovers" as a fraction.    I used to call that Choice #3

```
    0.16 R4
6) 1.00
   - 6
   ---
     40
   - 36
   ---
      4
```

$\frac{1}{6} = 0.16\frac{4}{6} = 0.16⅔$

✓ Use a bar over the repeating part.    I used to call that Choice #4

```
    0.1666666
6) 1.0000000
   - 6
   ---
     40
   - 36
   ---
     40
    etc.
```

$\frac{1}{6} = 0.16\overline{6}$

B) as a percent

$\frac{1}{6} = 16⅔\%$ or $16.\overline{6}\%$


872. Darlene knew exactly how Joe would spend his time after they got married.  Each day . . .

  12 hours at work
   1 hour at the gym
   8 hours sleeping
   3 hours cleaning the mansion

Draw a pie chart (circle graph).

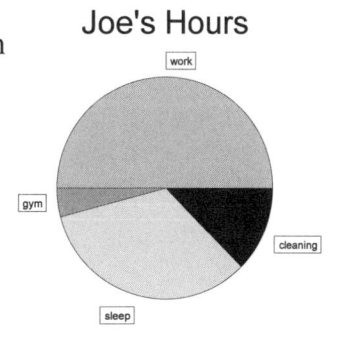

Joe's Hours

165

| 888–888 | Complete Solutions and Answers |

888. In the base 2 system (also known as the binary system)
1011 means $1 \times 2 \times 2 \times 2 + 0 \times 2 \times 2 + 1 \times 2 + 1 \times 1$.
$11101_{base\ 2}$ is equal to what in our base 10 system?

$11101_{base\ 2} = 1 \times 2 \times 2 \times 2 \times 2 + 1 \times 2 \times 2 \times 2 + 1 \times 2 \times 2 + 0 \times 2 + 1 \times 1$
$= 16 + 8 + 4 + 1 = 29_{base\ 10}$

The binary system is the one that computers feel comfortable with.

Electricity can either be flowing in a wire or not.

flowing = on = 1 in the binary system.
not flowing = off = 0 in the binary system.

Multiplication is really easy.
$0 \times 0 = 0$
$0 \times 1 = 0$
$1 \times 0 = 0$
$1 \times 1 = 1$

That's it! There is no having to memorize $7 \times 8 = 56$.

small essay
**Why Don't We Use Base 2?**

There are only two digits in the base 2 system: 0 and 1. The addition and multiplication tables are duck soup (easy as falling off a log). You never have to use complicated digits such as 2 or 3 or 4.

Hand-held calculators would look like this:

When we thought of using the base 20 system, the numbers were shorter, but the multiplication tables became huge.

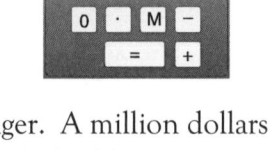

With the binary system the multiplication tables are super easy, but the numbers become longer. A million dollars ($1,000,000$_{base\ 10}$) in base 2 is $11,110,100,001,001,000,000$_{base\ 2}$.

Calculators that use the binary system would have to have tiny digits in the display.

# Complete Solutions and Answers | 890–896

➻ Using the binary (base 2) system would make the numbers very large.

➻ Using the vigesimal (base 20) system would make the multiplication tables very large.

In the words of Goldilocks, using our base 10 is just right—not too cold and not too hot.

*end of small essay*

890. Joe's future pet, Jelly Bean, weighs 1,800 pounds. One small jelly bean weighs 0.4 ounces. (1 pound = 16 ounces)

What is their difference in weight?

```
  1800 pounds                   1799 pounds   16 ounces
−            0.4 ounces       −             0.4 ounces
                                1799 pounds   15.6 ounces
```

Jelly Bean weighs 1,799 pounds 15.6 ounces more than a small jelly bean.

896. Which of these appears to be true?
  I) Primes get more plentiful as you go to larger and larger numbers.
  II) Primes seem to be equally scattered among all the numbers.
  III) Primes seem to get rarer as you go to larger and larger numbers.

As we look at that list of primes, we notice that . . .
☞ There are 8 primes between 1 and 20.
☞ There are 5 primes between 101 and 120.
☞ There are 3 primes between 901 and 920.

Primes seem to get rarer as you go to larger and larger numbers.

In fact, when we get to really large natural numbers, there are places where we have a million composite numbers in a row.

167

| 900–915 | Complete Solutions and Answers |

900. He filled his backpack with jelly beans from his 100-pound sack. After filling his backpack, his 100-pound sack now weighed 43.7 pounds. How many pounds of jelly beans had Joe transferred?

$$\begin{array}{r} 100 \\ -\phantom{0}43.7 \\ \hline \end{array} \qquad \begin{array}{r} 100.0 \\ -\phantom{0}43.7 \\ \hline 56.3 \end{array}$$

His backpack now had 56.3 pounds of jelly beans in it.

908. Change 0.33 into a fraction.

$0.33 = \frac{33}{100}$     How pleasant! Your baby brother could do this kind of problem all day long.

✔ $0.27 = \frac{27}{100}$

✔ $0.359 = \frac{359}{1000}$

✔ $0.9 = \frac{9}{10}$

✔ $0.88 = \frac{88}{100}$     The only thing you would have to help him with is reducing some of the answers. $\frac{88}{100} = \frac{44}{50} = \frac{22}{25}$

915. Yes. Each member of the domain has exactly one answer.

★ has one answer. ✦ has one answer. ☎ has one answer.

To determine if a rule is a function you look at the domain (the starting set). As Joe said, you have to be able to count up to one. You make sure that each member of the domain has exactly one (not zero and not two) arrows coming out of it. Then each member of the domain is assigned to exactly one answer.

# Complete Solutions and Answers | 930–944

**930.** Owning 5 dogs would cost $17.45 to feed them. Darlene thought it would be more fun to have 6 dogs. How much would it cost to feed 6 dogs?

Step 1: You know that 5 dogs = $17.45.

Step 2: The conversion factor will be either $\dfrac{5 \text{ dogs}}{\$17.45}$ or $\dfrac{\$17.45}{5 \text{ dogs}}$

Step 3: You are given 6 dogs. You want to convert that into dollars.

Step 4: You pick the conversion factor (from step 2) so that the units will cancel.

$$\frac{6 \text{ dogs}}{1} \times \frac{\$17.45}{5 \text{ dogs}} = \frac{6 \text{ dogs}}{1} \times \frac{\$17.45}{5 \text{ dogs}} =$$
$$6 \times \$17.45 \div 5 = \$104.70 \div 5 = \$20.94$$

It will cost $20.94 per week to feed six dogs.

What the article in BRIDAL DOGS magazine didn't mention was the additional costs of vet bills and dog licenses. It also didn't point out that some apartment leases don't allow dogs. If Darlene got six dogs, she would have to move.

**944.** The box would measure 10.8' × 18.' How many square feet will that be?

```
    10.8        1 digit to the right of the decimal
  ×   18        0 digits to the right of the decimal
    1944   ⇒   194.4 square feet.
```

Darlene explained to Joe that they were going to have so many cats that they would have to hire "catboys" (cowboys for cats).

Had Joe been paying attention, he would have been frightened.

Not much would frighten Joe because he had the attention span of a goldfish.

## 962–977 Complete Solutions and Answers

**962.** The first example is not a function. 5 is in the domain, but it doesn't have an answer. Each member of the domain must have an answer.

The second example is a function. Each member of the domain has exactly one answer.

The third example is not a function. 1 is in the domain and it has two answers. $1 \to 1$ and $1 \to 3$.

**965.** Change 87½% into a fraction.

$$87\tfrac{1}{2}\% = \frac{87\tfrac{1}{2}}{100}$$

$$= 87\tfrac{1}{2} \div 100$$

$$= \frac{175}{2} \times \frac{1}{100}$$

$$= \frac{175}{200} = \frac{7}{8}$$

**977.** The hole had a diameter of 2.5 feet. What is the circumference? (Use 3.1 for $\pi$.)

$C = \pi d \approx 3.1 \times 2.5$

```
      3.1     one digit to the right of the decimal
   ×  2.5     one digit to the right of the decimal
     155
     62
     775  ⟹ 7.75   two digits to the right of the decimal
```

Darlene will need 7.75 feet of ribbon to go around the hole.

# Index

adding decimals
- #130. . . . . . . . . . . . . . . . . 6
- #150. . . . . . . . . . . . . . . . . 7
- #250. . . . . . . . . . . . . . . . . 6
- #346. . . . . . . . . . . . . . . . 44
- #380. . . . . . . . . . . . . . . . . 6
- #525. . . . . . . . . . . . . . . . 11
- #567. . . . . . . . . . . . . . . . . 6
- #600. . . . . . . . . . . . . . . . 19
- #656. . . . . . . . . . . . . . . . 14
- #690. . . . . . . . . . . . . . . . . 6
- #720. . . . . . . . . . . . . . . . . 6

area and volume
- #139. . . . . . . . . . . . . . . . 57
- #141. . . . . . . . . . . . . . . . 61
- #172. . . . . . . . . . . . . . . . 63
- #200. . . . . . . . . . . . . . . . 64
- #231. . . . . . . . . . . . . . . . 73
- #268. . . . . . . . . . . . . . . . 63
- #285. . . . . . . . . . . . . . . . 61
- #307. . . . . . . . . . . . . . . . 61
- #348. . . . . . . . . . . . . . . . 63
- #393. . . . . . . . . . . . . . . . 47
- #405. . . . . . . . . . . . . . . . 68
- #442. . . . . . . . . . . . . . . . 74
- #446. . . . . . . . . . . . . . . . 47
- #484. . . . . . . . . . . . . . . . 49
- #569. . . . . . . . . . . . . . . . 63
- #590. . . . . . . . . . . . . . . . 52
- #641. . . . . . . . . . . . . . . . 71
- #688. . . . . . . . . . . . . . . . 68
- #838. . . . . . . . . . . . . . . . 50

bar graphs
- #168. . . . . . . . . . . . . . . . 36
- #344. . . . . . . . . . . . . . . . 36
- #425. . . . . . . . . . . . . . . . 44
- #588. . . . . . . . . . . . . . . . 36
- #615. . . . . . . . . . . . . . . . 55
- #798. . . . . . . . . . . . . . . . 62

conversion factors
- #126. . . . . . . . . . . . . . . . 35
- #137. . . . . . . . . . . . . . . . 52
- #234. . . . . . . . . . . . . . . . 33
- #236. . . . . . . . . . . . . . . . 57
- #274. . . . . . . . . . . . . . . . 44
- #301. . . . . . . . . . . . . . . . 73
- #310. . . . . . . . . . . . . . . . 35
- #327. . . . . . . . . . . . . . . . 19
- #399. . . . . . . . . . . . . . . . 35
- #406. . . . . . . . . . . . . . . . 21
- #429. . . . . . . . . . . . . . . . 64
- #440. . . . . . . . . . . . . . . . 19
- #448. . . . . . . . . . . . . . . . 62
- #473. . . . . . . . . . . . . . . . 42
- #511. . . . . . . . . . . . . . . . 80
- #542. . . . . . . . . . . . . . . . 68
- #549. . . . . . . . . . . . . . . . 31
- #577. . . . . . . . . . . . . . . . 24
- #630. . . . . . . . . . . . . . . . 45
- #643. . . . . . . . . . . . . . . . 26
- #649. . . . . . . . . . . . . . . . 62
- #658. . . . . . . . . . . . . . . . 47
- #714. . . . . . . . . . . . . . . . 56
- #810. . . . . . . . . . . . . . . . 16
- #930. . . . . . . . . . . . . . . . 39

## Index

d = rt  distance, rate, time
- #158. . . . . . . . . . . . . . . . . . . 56
- #214. . . . . . . . . . . . . . . . . . . 70
- #223. . . . . . . . . . . . . . . . . . . 34
- #278. . . . . . . . . . . . . . . . . . . 34
- #395. . . . . . . . . . . . . . . . . . . 56
- #475. . . . . . . . . . . . . . . . . . . 56
- #530. . . . . . . . . . . . . . . . . . . 66
- #557. . . . . . . . . . . . . . . . . . . 56
- #579. . . . . . . . . . . . . . . . . . . 29
- #677. . . . . . . . . . . . . . . . . . . 56
- #806. . . . . . . . . . . . . . . . . . . 45

decimals to fractions
- #128. . . . . . . . . . . . . . . . . . . 31
- #400. . . . . . . . . . . . . . . . . . . . 8
- #640. . . . . . . . . . . . . . . . . . . 11
- #750. . . . . . . . . . . . . . . . . . . . 8
- #908. . . . . . . . . . . . . . . . . . . 33

diameter, radius, circumference
- #102. . . . . . . . . . . . . . . . . . . 34
- #140. . . . . . . . . . . . . . . . . . . 49
- #155. . . . . . . . . . . . . . . . . . . 12
- #162. . . . . . . . . . . . . . . . . . . 34
- #191. . . . . . . . . . . . . . . . . . . 28
- #260. . . . . . . . . . . . . . . . . . . 12
- #360. . . . . . . . . . . . . . . . . . . 26
- #470. . . . . . . . . . . . . . . . . . . 12
- #504. . . . . . . . . . . . . . . . . . . 44
- #585. . . . . . . . . . . . . . . . . . . 16
- #602. . . . . . . . . . . . . . . . . . . 33
- #821. . . . . . . . . . . . . . . . . . . 62
- #977. . . . . . . . . . . . . . . . . . . 21

dividing decimals
- #191. . . . . . . . . . . . . . . . . . . 27
- #193. . . . . . . . . . . . . . . . . . . 37
- #209. . . . . . . . . . . . . . . . . . . 30
- #266. . . . . . . . . . . . . . . . . . . 27
- #304. . . . . . . . . . . . . . . . . . . 30
- #376. . . . . . . . . . . . . . . . . . . 27
- #383. . . . . . . . . . . . . . . . . . . 51
- #421. . . . . . . . . . . . . . . . . . . 30
- #494. . . . . . . . . . . . . . . . . . . 28
- #516. . . . . . . . . . . . . . . . . . . 31

divisibility rules
- #100. . . . . . . . . . . . . . . . . . . 28
- #276. . . . . . . . . . . . . . . . . . . 25
- #337. . . . . . . . . . . . . . . . . . . 55
- #356. . . . . . . . . . . . . . . . . . . 25
- #367. . . . . . . . . . . . . . . . . . . 66
- #408. . . . . . . . . . . . . . . . . . . 25
- #444. . . . . . . . . . . . . . . . . . . 31
- #519. . . . . . . . . . . . . . . . . . . 25
- #738. . . . . . . . . . . . . . . . . . . 25
- #854. . . . . . . . . . . . . . . . . . . 35

dollars and cents
- #186. . . . . . . . . . . . . . . . . . . 46
- #264. . . . . . . . . . . . . . . . . . . 46
- #365. . . . . . . . . . . . . . . . . . . 46
- #872. . . . . . . . . . . . . . . . . . . 68

elapsed time
- #239. . . . . . . . . . . . . . . . . . . 77
- #259. . . . . . . . . . . . . . . . . . . 77
- #269. . . . . . . . . . . . . . . . . . . 77
- #354. . . . . . . . . . . . . . . . . . . 77
- #570. . . . . . . . . . . . . . . . . . . 80

## Index

fractions to decimals
- #121. . . . . . . . . . . . . . . . . 30
- #184. . . . . . . . . . . . . . . . . 32
- #256. . . . . . . . . . . . . . . . . 32
- #341. . . . . . . . . . . . . . . . . 32
- #398. . . . . . . . . . . . . . . . . 11
- #478. . . . . . . . . . . . . . . . . 66
- #538. . . . . . . . . . . . . . . . . 55
- #591. . . . . . . . . . . . . . . . . 32
- #647. . . . . . . . . . . . . . . . . 32
- #740. . . . . . . . . . . . . . . . . 33
- #754. . . . . . . . . . . . . . . . . 66
- #844. . . . . . . . . . . . . . . . . 33
- #870. . . . . . . . . . . . . . . . . 52

fractions to percents
- #237. . . . . . . . . . . . . . . . . 75
- #319. . . . . . . . . . . . . . . . . 75
- #553. . . . . . . . . . . . . . . . . 48
- #870. . . . . . . . . . . . . . . . . 52

functions
- #110. . . . . . . . . . . . . . . . . 17
- #114. . . . . . . . . . . . . . . . . 21
- #156. . . . . . . . . . . . . . . . . 69
- #206. . . . . . . . . . . . . . . . . 69
- #207. . . . . . . . . . . . . . . . . 24
- #253. . . . . . . . . . . . . . . . . 72
- #306. . . . . . . . . . . . . . . . . 24
- #329. . . . . . . . . . . . . . . . . 69
- #340. . . . . . . . . . . . . . . . . 17
- #352. . . . . . . . . . . . . . . . . 49
- #456. . . . . . . . . . . . . . . . . 17
- #466. . . . . . . . . . . . . . . . . 76
- #515. . . . . . . . . . . . . . . . . 76
- #523. . . . . . . . . . . . . . . . . 64

- #528. . . . . . . . . . . . . . . . . 17
- #536. . . . . . . . . . . . . . . . . 35
- #547. . . . . . . . . . . . . . . . . 39
- #611. . . . . . . . . . . . . . . . . 35
- #624. . . . . . . . . . . . . . . . . 69
- #662. . . . . . . . . . . . . . . . . 17
- #668. . . . . . . . . . . . . . . . . 76
- #672. . . . . . . . . . . . . . . . . 80
- #680. . . . . . . . . . . . . . . . . 39
- #727. . . . . . . . . . . . . . . . . 39
- #759. . . . . . . . . . . . . . . . . 18
- #819. . . . . . . . . . . . . . . . . 35
- #836. . . . . . . . . . . . . . . . . 18
- #860. . . . . . . . . . . . . . . . . 37
- #915. . . . . . . . . . . . . . . . . 18
- #962. . . . . . . . . . . . . . . . . 18

Goldbach conjectures
- #104. . . . . . . . . . . . . . . . . 40
- #134. . . . . . . . . . . . . . . . . 42
- #215. . . . . . . . . . . . . . . . . 40
- #228. . . . . . . . . . . . . . . . . 42
- #308. . . . . . . . . . . . . . . . . 40
- #312. . . . . . . . . . . . . . . . . 42
- #313. . . . . . . . . . . . . . . . . 68
- #423. . . . . . . . . . . . . . . . . 72

graphing
- #273. . . . . . . . . . . . . . . . . 78
- #358. . . . . . . . . . . . . . . . . 72
- #595. . . . . . . . . . . . . . . . . 72
- #668. . . . . . . . . . . . . . . . . 76
- #694. . . . . . . . . . . . . . . . . 72

logic
- #607. . . . . . . . . . . . . . . . . 66
- #620. . . . . . . . . . . . . . . . . 30

## Index

minutes, hours, inches, feet
    #261. . . . . . . . . . . . . . . . . . 56
    #613. . . . . . . . . . . . . . . . . . 42
    #705. . . . . . . . . . . . . . . . . . 26

multiplying by ten, hundred, etc.
    #123. . . . . . . . . . . . . . . . . . 10
    #225. . . . . . . . . . . . . . . . . . 10
    #227. . . . . . . . . . . . . . . . . . 26
    #244. . . . . . . . . . . . . . . . . . 52
    #339. . . . . . . . . . . . . . . . . . 10
    #378. . . . . . . . . . . . . . . . . . 37
    #411. . . . . . . . . . . . . . . . . . 10
    #700. . . . . . . . . . . . . . . . . . 16

multiplying by zero
    #746. . . . . . . . . . . . . . . . . . 19

multiplying decimals
    #180. . . . . . . . . . . . . . . . . . 15
    #270. . . . . . . . . . . . . . . . . . 21
    #390. . . . . . . . . . . . . . . . . . 15
    #436. . . . . . . . . . . . . . . . . . 15
    #521. . . . . . . . . . . . . . . . . . 37
    #533. . . . . . . . . . . . . . . . . . 15
    #785. . . . . . . . . . . . . . . . . . 15
    #944. . . . . . . . . . . . . . . . . . 15

number systems
    #108. . . . . . . . . . . . . . . . . . . 5
    #119. . . . . . . . . . . . . . . . . . . 5
    #132. . . . . . . . . . . . . . . . . . 47
    #160. . . . . . . . . . . . . . . . . . . 9
    #166. . . . . . . . . . . . . . . . . . 26
    #212. . . . . . . . . . . . . . . . . . 39
    #222. . . . . . . . . . . . . . . . . . . 5
    #230. . . . . . . . . . . . . . . . . . 14
    #240. . . . . . . . . . . . . . . . . . . 7

    #247. . . . . . . . . . . . . . . . . . 47
    #258. . . . . . . . . . . . . . . . . . 39
    #280. . . . . . . . . . . . . . . . . . 11
    #325. . . . . . . . . . . . . . . . . . . 5
    #330. . . . . . . . . . . . . . . . . . . 9
    #333. . . . . . . . . . . . . . . . . . . 7
    #480. . . . . . . . . . . . . . . . . . . 5
    #624. . . . . . . . . . . . . . . . . . 69
    #764. . . . . . . . . . . . . . . . . . 21
    #766. . . . . . . . . . . . . . . . . . 62
    #852. . . . . . . . . . . . . . . . . . 11
    #888. . . . . . . . . . . . . . . . . . . 7

ordered pairs
    #156. . . . . . . . . . . . . . . . . . 69
    #206. . . . . . . . . . . . . . . . . . 69
    #253. . . . . . . . . . . . . . . . . . 72
    #273. . . . . . . . . . . . . . . . . . 78
    #329. . . . . . . . . . . . . . . . . . 69
    #358. . . . . . . . . . . . . . . . . . 72
    #423. . . . . . . . . . . . . . . . . . 72
    #466. . . . . . . . . . . . . . . . . . 76
    #515. . . . . . . . . . . . . . . . . . 76
    #595. . . . . . . . . . . . . . . . . . 72

percent problems
    #142. . . . . . . . . . . . . . . . . . 59
    #146. . . . . . . . . . . . . . . . . . 64
    #153. . . . . . . . . . . . . . . . . . 53
    #174. . . . . . . . . . . . . . . . . . 51
    #176. . . . . . . . . . . . . . . . . . 65
    #179. . . . . . . . . . . . . . . . . . 70
    #194. . . . . . . . . . . . . . . . . . 70
    #216. . . . . . . . . . . . . . . . . . 59
    #218. . . . . . . . . . . . . . . . . . 65
    #220. . . . . . . . . . . . . . . . . . 51

## Index

#263. . . . . . . . . . . . . . . . . 76
#282. . . . . . . . . . . . . . . . . 53
#321. . . . . . . . . . . . . . . . . 54
#323. . . . . . . . . . . . . . . . . 65
#335. . . . . . . . . . . . . . . . . 78
#342. . . . . . . . . . . . . . . . . 71
#357. . . . . . . . . . . . . . . . . 78
#362. . . . . . . . . . . . . . . . . 59
#388. . . . . . . . . . . . . . . . . 60
#410. . . . . . . . . . . . . . . . . 57
#471. . . . . . . . . . . . . . . . . 80
#492. . . . . . . . . . . . . . . . . 65
#496. . . . . . . . . . . . . . . . . 51
#506. . . . . . . . . . . . . . . . . 60
#517. . . . . . . . . . . . . . . . . 51
#532. . . . . . . . . . . . . . . . . 80
#593. . . . . . . . . . . . . . . . . 65
#685. . . . . . . . . . . . . . . . . 51
#846. . . . . . . . . . . . . . . . . 66

percents to decimals or fractions
#210. . . . . . . . . . . . . . . . . 48
#404. . . . . . . . . . . . . . . . . 48
#461. . . . . . . . . . . . . . . . . 75
#606. . . . . . . . . . . . . . . . . 50
#965. . . . . . . . . . . . . . . . . 76

pi
#102. . . . . . . . . . . . . . . . . 34
#199. . . . . . . . . . . . . . . . . 28
#246. . . . . . . . . . . . . . . . . 19
#292. . . . . . . . . . . . . . . . . 16
#360. . . . . . . . . . . . . . . . . 26
#585. . . . . . . . . . . . . . . . . 16
#602. . . . . . . . . . . . . . . . . 33
#977. . . . . . . . . . . . . . . . . 21

pie charts
#136. . . . . . . . . . . . . . . . . 48
#317. . . . . . . . . . . . . . . . . 57
#645. . . . . . . . . . . . . . . . . 48
#731. . . . . . . . . . . . . . . . . 50
#872. . . . . . . . . . . . . . . . . 68

pirme numbers
#107. . . . . . . . . . . . . . . . . 38
#205. . . . . . . . . . . . . . . . . 38
#302. . . . . . . . . . . . . . . . . 38
#419. . . . . . . . . . . . . . . . . 38
#502. . . . . . . . . . . . . . . . . 38
#761. . . . . . . . . . . . . . . . . 55
#813. . . . . . . . . . . . . . . . . 55
#896. . . . . . . . . . . . . . . . . 55

probability and fractions
#368. . . . . . . . . . . . . . . . . 79
#384. . . . . . . . . . . . . . . . . 79
#403. . . . . . . . . . . . . . . . . 79
#417. . . . . . . . . . . . . . . . . 79
#430. . . . . . . . . . . . . . . . . 79
#455. . . . . . . . . . . . . . . . . 79
#462. . . . . . . . . . . . . . . . . 80
#557. . . . . . . . . . . . . . . . . 56
#602. . . . . . . . . . . . . . . . . 33
#840. . . . . . . . . . . . . . . . . . 9

ratio
#147. . . . . . . . . . . . . . . . . 67
#272. . . . . . . . . . . . . . . . . 67
#381. . . . . . . . . . . . . . . . . 76
#385. . . . . . . . . . . . . . . . . 67
#413. . . . . . . . . . . . . . . . . 67

## Index

rounding numbers
- #246. . . . . . . . . . . . . . . . . . 19
- #292. . . . . . . . . . . . . . . . . . 16
- #334. . . . . . . . . . . . . . . . . . 21
- #336. . . . . . . . . . . . . . . . . . 14
- #402. . . . . . . . . . . . . . . . . . 26
- #454. . . . . . . . . . . . . . . . . . 39
- #490. . . . . . . . . . . . . . . . . . 24
- #531. . . . . . . . . . . . . . . . . . 47
- #532. . . . . . . . . . . . . . . . . . 80
- #538. . . . . . . . . . . . . . . . . . 55
- #581. . . . . . . . . . . . . . . . . . 13
- #591. . . . . . . . . . . . . . . . . . 32
- #663. . . . . . . . . . . . . . . . . . 37
- #677. . . . . . . . . . . . . . . . . . 56
- #685. . . . . . . . . . . . . . . . . . 51
- #701. . . . . . . . . . . . . . . . . . 31
- #711. . . . . . . . . . . . . . . . . . 45

sets
- #144. . . . . . . . . . . . . . . . . . 60
- #164. . . . . . . . . . . . . . . . . . 23
- #167. . . . . . . . . . . . . . . . . . 68
- #241. . . . . . . . . . . . . . . . . . 66
- #254. . . . . . . . . . . . . . . . . . 23
- #337. . . . . . . . . . . . . . . . . . 55
- #350. . . . . . . . . . . . . . . . . . 23
- #356. . . . . . . . . . . . . . . . . . 25
- #367. . . . . . . . . . . . . . . . . . 66
- #386. . . . . . . . . . . . . . . . . . 31
- #452. . . . . . . . . . . . . . . . . . 33
- #458. . . . . . . . . . . . . . . . . . 55
- #463. . . . . . . . . . . . . . . . . . 23
- #500. . . . . . . . . . . . . . . . . . 33
- #513. . . . . . . . . . . . . . . . . . 23
- #540. . . . . . . . . . . . . . . . . . 26
- #609. . . . . . . . . . . . . . . . . . 23
- #673. . . . . . . . . . . . . . . . . . 23
- #709. . . . . . . . . . . . . . . . . . 37
- #858. . . . . . . . . . . . . . . . . . 39

squares and circles
- #298. . . . . . . . . . . . . . . . . . 49
- #590. . . . . . . . . . . . . . . . . . 52

subtracting decimals
- #117. . . . . . . . . . . . . . . . . . . 8
- #170. . . . . . . . . . . . . . . . . . 11
- #178. . . . . . . . . . . . . . . . . . 66
- #242. . . . . . . . . . . . . . . . . . . 8
- #316. . . . . . . . . . . . . . . . . . 78
- #416. . . . . . . . . . . . . . . . . . 37
- #508. . . . . . . . . . . . . . . . . . . 9
- #586. . . . . . . . . . . . . . . . . . 42
- #665. . . . . . . . . . . . . . . . . . 52
- #678. . . . . . . . . . . . . . . . . . 19
- #697. . . . . . . . . . . . . . . . . . . 9
- #777. . . . . . . . . . . . . . . . . . 11
- #815. . . . . . . . . . . . . . . . . . 29
- #830. . . . . . . . . . . . . . . . . . . 8
- #900. . . . . . . . . . . . . . . . . . . 8

subtracting mixed units
- #125. . . . . . . . . . . . . . . . . . 24
- #238. . . . . . . . . . . . . . . . . . 20
- #300. . . . . . . . . . . . . . . . . . 20
- #450. . . . . . . . . . . . . . . . . . 20
- #544. . . . . . . . . . . . . . . . . . 20
- #651. . . . . . . . . . . . . . . . . . 20
- #670. . . . . . . . . . . . . . . . . . 24
- #800. . . . . . . . . . . . . . . . . . 31
- #890. . . . . . . . . . . . . . . . . . 21